Machine Learning-based Design and Optimization of High-Speed Circuits

Vazgen Melikyan

Machine Learning-based Design and Optimization of High-Speed Circuits

Vazgen Melikyan
Synopsys Armenia CJSC
Yerevan, Armenia

ISBN 978-3-031-50716-8   ISBN 978-3-031-50714-4 (eBook)
https://doi.org/10.1007/978-3-031-50714-4

© The Editor(s) (if applicable) and The Author(s), under exclusive license to Springer Nature Switzerland AG 2024

This work is subject to copyright. All rights are solely and exclusively licensed by the Publisher, whether the whole or part of the material is concerned, specifically the rights of translation, reprinting, reuse of illustrations, recitation, broadcasting, reproduction on microfilms or in any other physical way, and transmission or information storage and retrieval, electronic adaptation, computer software, or by similar or dissimilar methodology now known or hereafter developed.

The use of general descriptive names, registered names, trademarks, service marks, etc. in this publication does not imply, even in the absence of a specific statement, that such names are exempt from the relevant protective laws and regulations and therefore free for general use.

The publisher, the authors, and the editors are safe to assume that the advice and information in this book are believed to be true and accurate at the date of publication. Neither the publisher nor the authors or the editors give a warranty, expressed or implied, with respect to the material contained herein or for any errors or omissions that may have been made. The publisher remains neutral with regard to jurisdictional claims in published maps and institutional affiliations.

This Springer imprint is published by the registered company Springer Nature Switzerland AG
The registered company address is: Gewerbestrasse 11, 6330 Cham, Switzerland

Paper in this product is recyclable.

Dedicated to the bright memory of my parents:
Shavarsh Melikyan and Siranush Hakobyan

Preface

The book systematically expounds the main results obtained by the author in the field of design and optimization of high-speed integrated circuits (ICs) and their standard blocks (heterogeneous ICs, analog-to-digital and digital-to-analog converters, input/output cells, etc.) operating in non-standard conditions (deviations of technological process parameters, supply voltage, ambient temperature, etc.). The proposed methods are based on machine learning and consider effects of different external and internal destabilizing factors (radiation exposure, self-heating, nonideality of the power source, input signals, interconnects, power rails, etc.).

The main goals of most proposed methods and solutions of design and optimization of high-speed ICs are to improve important parameters and characteristics (performance, power consumption, occupied area on the die, transmitting and receiving data quality and accuracy) of circuits and reduce the effects of non-standard operating conditions and different types of destabilizing factors.

The book is anticipated for scientists and engineers specializing in the field of IC design and optimization as well as for students and postgraduate students studying disciplines related to IC design.

Yerevan, Armenia   Vazgen Melikyan

Introduction

Over the past 60 years, the semiconductor industry has evolved according to Moore's law, which has allowed to have metal-oxide-semiconductor (MOS) transistors or their different modifications (FinFET, GAA (Gate All Around), etc.) with channel lengths up to several nanometers. Leading companies/manufacturers, such as Taiwan Semiconductor Manufacturing Company (TSMC), Global Foundries (GF), and Samsung, have already developed 7, 5, and 3-nanometer technologies of IC fabrication. In the coming years, it is even possible to move to transistors with several Angstrom sizes. At present, already produced ICs contain more than hundred billions of active devices in ICs. Taking the planned transition to several Angstrom size technologies in the near future into account, the number of circuit components will soon increase significantly and can reach up to trillion. Such kind of fast developments have led to the enormous changes of the main parameters and characteristics of ICs.

Fast developments in semiconductor industry first affected the performance of ICs. The productivity of ICs drastically increased because the small size of transistors decreases their parasitic parameters and allows to have more functional blocks on the same area. That is why the operating frequencies of ICs have also increased, reaching dozens of GHz.

In addition to high-speed operation of contemporary ICs, the speed of transmission of information between them is also growing drastically. Modern electronic systems include different ICs, which are located independently of each other in the same Printed Circuit Board (PCB) and perform various functions. These ICs are in constant communication with each other and exchange processed information. Significant increase in the volume of transmitted data between different ICs dictates the necessity of high-speed transmission of information between them. Currently, the speed reaches up to 256 and 512 Gbit/s. For accurate operation of the entire electronic system, it is necessary to ensure correct transmission of information between different ICs of the system. Therefore, there is need to design special basic blocks in ICs that will ensure transmission and reception of information throughout the system. Such function is performed by specially designed input/

output (I/O) cells implemented in ICs. I/O cells are one of the most important components of contemporary ICs. The new standards of information transmission by I/O cells are aimed at increasing speed. For example, 4th generation Universal Serial Bus (USB) devices provide 40 Gbps data transfer speed, while 1st generation USB standard provided only 12 Mbit/s speed. Among the modern data transmission standards are Peripheral Component Interconnect (PCI) Express, Serial Advanced Technology Attachment (ATA), Double Data Rate (DDR), low-power DDR, High-Definition Multimedia Interface (HDMI), etc. These standards have many varieties that operate at different frequencies. They also differ from each other in their application. For example, DDR standard provides communication between computer core and external devices. Low-power DDR standard provides communication in modern smart phones. In the mentioned standards, information is transmitted sequentially using two transmission lines. Such choice of the number of transmission lines is related to the advantages of differential signal. It allows getting two-times increase in the signal level as well as avoiding the influence of common component of noise.

Development of high-speed ICs and their different blocks (in particular, I/Os) have led to strict requirements for IC reliability and need to develop means of designing ICs that work in non-standard operating conditions. Disruption of IC operation can directly or indirectly lead to serious economic losses, environmental damage, or threats to human life. The reduction in transistor sizes led to thinning of oxide layer of transistor gates which allowed reducing their threshold voltage. The low threshold voltage, in turn, allowed reducing supply voltages of ICs, leading to the reduction of power consumption. The development of technological process has made it possible to reduce the distance between metal layers, leading to an increase in the density of interconnects. In case of closer wires, the mutual capacitances between them increase. Large capacitances lead to an increase in the noise level. In modern ICs, the value of the supply voltage has become several hundred millivolts, and noises reach tens of millivolts. An increase in noise and a decrease in the value of the supply voltage led to a decrease in the signal-to-noise ratio. This reduces the noise immunity of modern ICs and can lead to unwanted changes in the characteristics of circuits, up to functional failure. The new generation I/O cells are faster and work with lower supply voltages, which significantly complicates the processes of data reading and transfer. It should be noted that as a result of reducing the supply voltage, ICs become sensitive to external and internal destabilizing factors, and due to the increase in speed, the transmission line begins to significantly suppress the transmitted data which makes data processing even more difficult. Currently, there are also high-precision IC types which provide extremely high reliability. They are mainly used in medical fields, space equipment, etc. Since the loss of data in the above areas can lead to great damage, I/O cells in high-precision ICs are equipped with systems to increase the reliability of data transmission and reading. With the increase of data transfer rates, deviations of timing parameters of data become an extremely important problem. In modern I/O cells, the fastest data change time reaches picoseconds. In that case, even small deviations of timing parameters can lead to data error. The main obstacles to increasing the speed of data transmitted in

I/O cells and their lossless transmission are process-voltage-temperature deviations, the influence of transmission lines, deviations caused in transmitter sub-connections, etc. The increase in operating frequencies, the density of transistors and interconnects per unit area, and the reduction of the thickness of the oxide layer have made the influence of undesirable phenomena of self-heating and aging on the operation of ICs more tangible. It leads to complications in the IC design process. The development of ICs and their different blocks operating in non-standard conditions becomes a critical issue. There is a need to develop new methods of designing ICs that operate under non-standard conditions, which will reduce self-heating and aging phenomena on the operation of circuits while increasing the noise immunity and efficiency.

Thus, an extreme need arises for the development of design and optimization methods of high-speed ICs and their separate blocks. For example, the design of means to restore the distorted signal has become extremely relevant at present, because due to the increase in the amount of data processed in high-speed systems, in order to meet the demand for increasing the speed of their transmission, the speed of reception of sequential information is limited by the amplitude-frequency characteristics of the channel. Another example is that, as a result of changes in external factors, the deviation of transistor parameters and the increase in the effect of aging phenomena have led to decrease in the efficiency of calibration nodes and equalizers included in ICs, which can even lead to functional failures. From another side, the development of design methods for ICs that will work under non-standard operating conditions has become an extremely important challenge. As an example, one of the most relevant problems in the creation of modern ICs is the development of signal transmission calibration means, which can significantly improve their main parameters and contribute to further increase in the speed of I/O cells.

The existing means of design and optimization of high-speed ICs are not able to solve the mentioned problems since they are oriented at larger technology nodes and other operating conditions. A series of reasons exist that make it difficult to apply the existing means and available methods of design and optimization of high-speed ICs in contemporary real IC design practice. For example, experimental studies of existing means of increasing the speed of receiving sequential information and the results of their analysis show that they do not completely solve the problems in ICs and have several time limitations that are unacceptable from the point of view of practical application. That circumstance dictates the need to develop more effective means of the mentioned class. The existing means of designing ICs working in non-standard operating conditions are based on calibration algorithms and are oriented at technological processes of larger sizes. In addition, the possibility of sudden changes in temperature and voltage after calibration and the increased influence of aging phenomena typical of modern technologies are not taken into account. This indicates that the latter do not meet modern requirements, and the need to develop new solutions to increase stability and reliability of ICs has arisen. Research of existing approaches and means of signal transmission systems in ICs shows that although they provide significant improvement in speed, data transmission and reading reliability, and power consumption, from the point of view of efficiency, the latter do not meet the modern requirements for practical design. Such kind of examples are too many.

The development of means of design and optimization of high-speed ICs has now become a decisive part in the process of IC design. The problems, solved during design and especially the optimization of high-speed ICs, require huge amount of computations because of enormous number of components and options of considering versions of designing circuits and their operating modes. From this point of view and taking into account current rise of machine learning (ML), connected with the occurrence of big data and more powerful computing resources in the last few years, ML methods and tools can be considered as most suitable during design and optimization of high-speed ICs.

This monograph is devoted to description of the developed new principles, methods, and circuit solutions for design and optimization of high-speed ICs, based on ML. In the monograph different effective new principles, methods, solutions, and means of design and optimization of high-speed ICs are proposed. For example, effective approaches from the point of view of timing limitations for the development of means to increase speed of sequential information reception of ICs have been proposed. The embedded nodes and the structure of their architecture allow to significantly increase data transfer and processing frequencies in case of increasing the area occupied on the die and power consumption within the permissible limits. Methods of IC design that work in non-standard operating conditions have been proposed which meet modern requirements and, at the expense of increasing the occupied area and power consumption within the permissible limits, significantly reduce the deviations caused by changes in external and internal destabilizing factors and aging phenomena. Principles and methods of developing signal transmission calibration means in ICs have been proposed, which allow to significantly improve their main technical characteristics and parameters: speed, reliability of data transfer and reading, power consumption.

The author expresses deep gratitude to his PhD students Manvel Grigoryan, Hakob Kostanyan, Karen Khachikyan, Arman Atanesyan, Taron Kaplanyan, and Suren Abazyan, who have participated in the development of described means.

The author would also like to thank his colleague Ruzanna Goroyan who assisted in the preparation of this monograph.

Contents

1 Means to Accelerate Transfer of Information Between Integrated Circuits........ 1
  1.1 General Issues of Means to Accelerate Transfer of Information Between Integrated Circuits........ 1
    1.1.1 Importance of Means to Accelerate Transfer of Information Between Integrated Circuits........ 1
    1.1.2 Current State and Issues of Existing Means of Increasing the Speed of Receiving Sequential Information in an Integrated Circuit........ 12
    1.1.3 Proposed Approaches to Increase the Speed of Receiving Sequential Information in an Integrated Circuit........ 25
  1.2 Methods to Increase the Speed of Receiving Sequential Information in an Integrated Circuit........ 28
    1.2.1 Method to Increase the Speed of Receiving Sequential Information with the Use of a Circuit with Negative Capacitance in Continuous Time Linear Equalizer (CTLE)........ 28
    1.2.2 Method to Increase the Speed of Receiving Sequential Information with the Use of a High-Speed Comparator with Low Input Capacitance in the Decision Feedback Equalizer (DFE)........ 40
    1.2.3 Method to Increase the Speed of Receiving Sequential Information Due to Processing of the Signal Transmitted by Pulse Amplitude Modulation (PAM4), Conditioned by Parallel Branches........ 46
  References........ 52

2	**Design Methods of Integrated Circuits, Working Under Non-standard Operating Conditions**............................		59
	2.1	General Issues of Design Methods of Integrated Circuits, Working Under Non-standard Operating Conditions...........	59
		2.1.1 Need for Design Methods of Integrated Circuits, Working Under Non-standard Operating Conditions.....	59
		2.1.2 Existing Design Methods of Integrated Circuits, Working Under Non-standard Operating Conditions.....	63
		2.1.3 Existing Problems of Design Methods of Integrated Circuits, Working Under Non-standard Operating Conditions..	76
		2.1.4 Principles of Increasing the Stability of Integrated Circuits, Working Under Non-standard Operating Conditions..	81
	2.2	Reduction of Offset Value Caused by aging Phenomena Through Adding Transmission Gates and Transistor Switches in Comparators................................	81
	2.3	Incorporation of the Current DAC in the Method of Offset Reduction Through the Circuit with Digital Control of IC Receiver......................................	82
	2.4	Implementation of Negative Feedback in DDLs..............	82
		2.4.1 Conclusions..................................	82
	2.5	Design Methods of Integrated Circuits, Working Under Non-standard Operating Conditions.......................	83
		2.5.1 Method of Reducing Offset, Caused by Aging Phenomena in Comparators.......................	84
		2.5.2 Method of Reducing the Offset Caused by Sharp Fluctuations in Ambient Temperature in the Receiver....	87
		2.5.3 Method of Reducing the Offset Caused by Sharp Voltage and Temperature Fluctuations in Digital Delay Lines....................................	93
	2.6	Conclusions...	101
	References...		102
3	**Signal Transmission Calibration Systems in Integrated Circuits**....		109
	3.1	General Issues of Signal Transmission Calibration Systems in Integrated Circuits.................................	109
		3.1.1 Importance of Signal Transmission Calibration Means in Integrated Circuits...........................	109
		3.1.2 Importance of Developing Signal Transmission Calibration Means in Integrated Circuits..............	113
		3.1.3 Current State and Issues of Design Methods of Signal Transmission Calibration Systems in Integrated Circuits..	121

	3.2	Design Principles of Signal Transmission Calibration Systems in Integrated Circuits.....................................	129
		3.2.1	Method of Detection and Self-Regulation of Duty Cycle Deviations in High-Speed Integrated Circuits.....	129
		3.2.2	Method of Increasing the Reliability of High-frequency Data in Transmitters................................	137
		3.2.3	Method of Calibration of Asymmetries of Rise/Fall Times of High-frequency Signals...................	144
		3.2.4	Calibration Method of Signal Distortion Caused by Transmission Line...................................	150
	References...			159
4	**Methods to Improve Linearity of Signal's Analog-to-Digital Conversion with Self-Calibration**.............................			165
	4.1	General Issues to Improve Linearity of Signal's Analog-to-Digital Conversion with Self-Calibration............................		165
		4.1.1	Importance of Means to Improve the Linearity of Signal's Analog-to-Digital Conversion with Self-Calibration................................	165
		4.1.2	Need for Means to Improve the Linearity of Signal's Analog-to-Digital Conversion with Self-Calibration................................	169
		4.1.3	Causes of Nonlinearity Occurrence in Signal's Analog-to-Digital Conversion and the Importance and Necessity of its Reduction in Flash ADCs..........	170
		4.1.4	Causes of Nonlinearity Occurrence Signal's in Analog-to-Digital Conversion and the Importance and Necessity of its Reduction in Current DACs........	173
		4.1.5	Causes of Nonlinearity Occurrence in Signal's Analog-to-Digital Conversion and the Importance and Necessity of Its Reduction in Pipeline ADCs........	176
		4.1.6	Existing Means to Improve the Linearity of Signal's Analog-to-Digital Conversion with Self-Calibration.....	182
		4.1.7	Existing Methods for Reducing Nonlinearity in Flash ADCs...	183
		4.1.8	Existing Methods for Reducing Nonlinearity in Current DACs...	186
		4.1.9	Existing Methods for Reducing Nonlinearity in Pipeline ADCs...	189
		4.1.10	Principles to Improve the Linearity of Signal's Analog-to-Digital Conversion with Self-Calibration.....	193
	4.2	Conclusions...		194
	4.3	Methods of Improving the Linearity of Signal's Analog-to-Digital Conversion with Self-Calibration............		195

		4.3.1	Method of Reducing the Nonlinearity with Self-Calibration by Correcting the Offset Error of Comparators in Flash Analog-to-Digital Converters	195
		4.3.2	Method of Reducing the Nonlinearity with Self-Calibration by Correcting the Current Deviation Error of Current Sources in Current Digital-to-Analog Converters	205
		4.3.3	Means of Reducing System Nonlinearity with Self-Calibration by Increasing the Linearity of Comparators in Pipeline Analog-to-Digital Converters	210
	4.4	Conclusion...		214
	References...			214
5	**Design of High-performance Heterogeneous Integrated Circuits**			221
	5.1	General Issues of Designing Means for High-performance Heterogeneous Integrated Circuits........................		221
		5.1.1	Importance of Design Means for High-performance Heterogeneous Integrated Circuits	221
		5.1.2	Current State and Issues of Design Means for High-performance Heterogeneous Integrated Circuits	230
		5.1.3	Principles of Design Means for High-performance Heterogeneous Integrated Circuits	241
	5.2	Methods of Design for High-performance Heterogeneous Integrated Circuits.......................................		242
		5.2.1	Method for Improving Data Transfer Between Components in High-performance Heterogeneous Integrated Circuits...............................	242
		5.2.2	Method for Improving Data Transfer Between Clock Domains in High-performance Heterogeneous Integrated Circuits.......................................	253
		5.2.3	Implementation Method of the Architecture in High-performance Heterogeneous Integrated Circuits.......................................	263
	References...			274
6	**Design of Digital Integrated Circuits by Improving the Characteristics of Digital Cells**..........................			279
	6.1	General Issues in Design of Digital Integrated Circuits by Improving the Characteristics of Digital Cells		279
		6.1.1	Importance of Design of Digital Integrated Circuits by Improving the Characteristics of Digital Cells	279
		6.1.2	Current State and Issues of Design of Digital Integrated Circuits by Improving the Characteristics of Digital Cells..	282

		6.1.3	Proposed Principles for Efficient Design of Integrated Circuits by Improving the Characteristics of Digital Cells.	304
	6.2		Methods of Design of Digital Integrated Circuits by Improving the Characteristics of Digital Cells.	308
		6.2.1	Method of Optimizing the Accessibility of Standard I/O Cells.	308
		6.2.2	Enhanced Method for I/O Cell Accessibility Prediction and Design Optimization with Proposed Machine Learning	312
		6.2.3	Optimization Method of Standard Cells for Digital Integrated Circuit Designs with Different Cell Heights.	318
		6.2.4	"Sleep Mode" Integration Method of Cells Using Neural Network for Low-Power Designs.	322
		6.2.5	Method of Adding Metal Fillers for Digital IC Design Flow.	325
References.				331
Index.				337

Chapter 1
Means to Accelerate Transfer of Information Between Integrated Circuits

1.1 General Issues of Means to Accelerate Transfer of Information Between Integrated Circuits

1.1.1 Importance of Means to Accelerate Transfer of Information Between Integrated Circuits

It is difficult to imagine the operation of modern electronic systems without integrated circuits (ICs), which are located independently of each other and perform various functions [1–5]. Despite this, ICs are in constant communication with each other and exchange processed information [6, 7]. Over time, the reduction of the minimum channel length of complementary metal-oxide-semiconductor (CMOS) transistors allows to include a larger number of components in an IC and increase their functions [8]. Thus, the amount of processed and transmitted data increases.

For the accurate operation of the entire system, it is necessary to ensure the correct transmission of information between different ICs of the system [9, 10]. Therefore, there is a need to design special basic working blocks in ICs that will ensure the transmission and reception of information throughout the system. Such a function is performed by specially designed input-output (I/O) cells implemented in ICs [11–13]. Currently, there are many structures of I/O cells that are used in various systems.

Over time, the continuous scaling of the channel of CMOS transistors has led to an increase in data transfer rates, as well as a decrease in supply voltages. Therefore, due to the I/O cells in ICs, the new standards of information transmission are aimed at increasing speed. For example, fourth-generation Universal Serial Bus (USB) devices provide 40 Gbps data transfer speed, while first-generation USB standard provided only 12 Mbit/s speed [14]. Among the modern data transmission standards, Peripheral Component Interconnect (PCI) Express, Serial Advanced Technology Attachment (ATA), double data rate (DDR), low-power DDR, and High-Definition Multimedia Interface (HDMI) standards are also notable [15–19]. The above standards also have many varieties that operate at different frequencies.

Data transmission standards also differ from each other in their application. For example, the DDR standard provides communication between the computer core and external devices, and the low-power DDR standard provides communication between modern smartphones.

In the above standards, information is transmitted sequentially using two transmission lines. Such a choice of the number of transmission lines is related to the advantages of the differential signal [20]. It allows to get a two-time increase in the signal level, as well as to avoid the influence of the common component of noises. Although there are special I/O cells that use the parallel method of data transfer [21–23], the sequential method is considered more preferable. This is due to the increase in the cost of the system as a result of adding additional channels during parallel transmission. The parallel transmission method also leads to an inevitable increase in the surface area of the IC due to the need for additional nodes for signal transmission, reception, and restoration.

Special blocks generating and processing a clock signal are an important part of special I/O cells performing the function of sequential information transmission and reception in ICs [24–26]. In general, three types of architectures with clock signal are distinguished in I/O cells: common, transmitted, and embedded [27–29].

In the case of a common clock signal architecture, there is one main quartz generator that provides exchange of reference clock signal to two functioning ICs (Fig. 1.1).

Such a structure requires providing equal-length interconnects to reduce the phase shift between the clock signals provided to two ICs. The architecture with a common clock signal allows to provide up to 100 Mb/s speed, so it is designed for low-power systems.

In the case of architecture with a transmitted clock signal, the transmitting IC provides a reference clock signal to the receiver (Fig. 1.2).

Necessity rises to add an additional routing between the transmitter and receiver nodes in the system to ensure the transmission of the clock signal. Such an

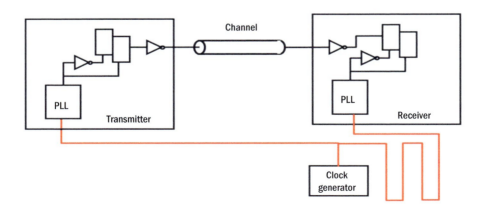

Fig. 1.1 Architecture with a common clock signal

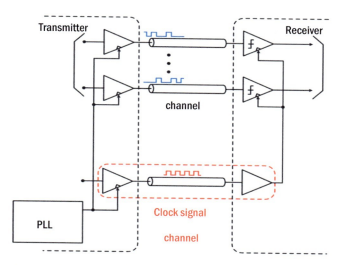

Fig. 1.2 Architecture with a transmitted clock signal

architecture is used in both processor-memory and multiprocessor communication interfaces. However, the disadvantage of the system is the limitation of the transmitted clock signal speed due to the frequency characteristics of the transmission channel. Since the channel has a characteristic similar to a low-pass filter, it significantly affects the transmitted clock signal by distorting its duty cycle and causes rise/falls to form at different times [30, 31]. There is also a need to design special fundamental blocks that will perform function of equalization for clock signal [32].

In the embedded *clock signal* architecture, there are basic blocks that provide reference clock signal and are present at both the receiver and transmitter (Fig. 1.3).

This architecture is used in "mesochronous" and "plesiochronous" systems [33, 34]. At the receiver, the frequency and optimal phase position of the clock signal are obtained from the input data stream. To perform the above function, special clock data recovery (CDR) blocks are designed, which can use the output signal of phase-locked loop (PLL) as a reference clock signal [35–40]. Such a structure provides higher performance and lower cost for the overall system, but increases the problems that rise during the design.

Constraints to Increase the Speed of Receiving Serial Information in an Integrated Circuit

I/O cells consist of the transmitter and receiver nodes as well as special cells that provide reference voltage and current and perform system setup (Fig. 1.4) [41–49].

The transmitter and receiver are connected by a channel. It includes all the components by which information is exchanged and is made up of wires and via connections (Fig. 1.5).

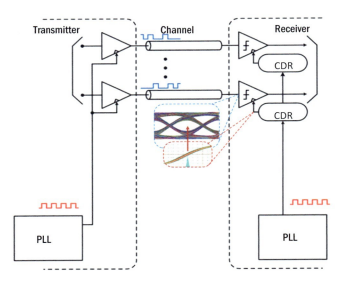

Fig. 1.3 Architecture with embedded clock signal

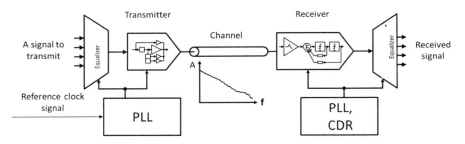

Fig. 1.4 Structure of I/O cell

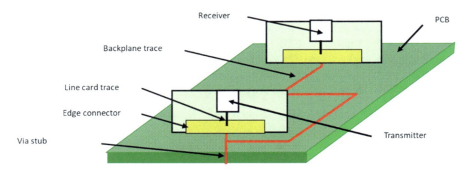

Fig. 1.5 Structure of a channel

1.1 General Issues of Means to Accelerate Transfer of Information... 5

Fig. 1.6 Microstrip and strip line structure of conductors in channels

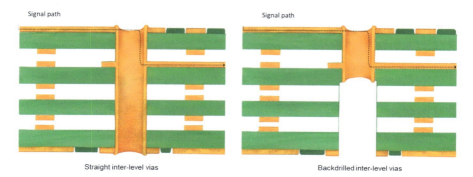

Fig. 1.7 Structure of straight and backdrilled inter-level vias

Conductors in channels can be of two types: microstrip and strip line (Fig. 1.6) [50, 51].

Although the strip line structure of wires is more complicated and implies higher costs during the manufacturing process, it allows to significantly avoid the noise caused by the effect of interconnects which is even more fundamental in the case of a differential signal.

There are also two types of vias located on the printed circuit board: straight and backdrilled (Fig. 1.7) [52–54].

Straight inter-level vias add extra zeros to the overall channel transfer function and increase the amount of signal distortion. This can be avoided if their isolated parts are removed at an additional stage during the manufacturing process. As a result, the cost of the system increases, but a more linear characteristic of the channel is obtained.

Increasing the speed of information exchange standards implemented by modern high-speed I/O cells makes the presentation format with centralized parameters of the channel impermissible [55–57], because such an approach does not take into account the following:

- Voltage drops in wires
- Changing magnetic field
- Bias currents
- Conduction currents due to dielectric imperfection

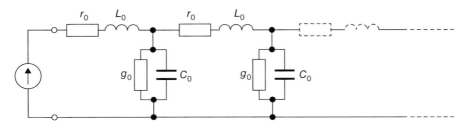

Fig. 1.8 A long line structure with distributed parameters

Therefore, increasing the frequency of the supply voltage is a necessary condition for presenting the channel as a long line with distributed parameters. It is made up of a large number of r_0, g_0, L_0, and C_0 components (Fig. 1.8).

R_0 and g_0 are conditioned by thermal and dielectric losses, and L_0 and C_0 components model the interactions of the electric and magnetic fields between the conductors. It is convenient to consider such lines, the above-mentioned primary parameters of which are constant along the entire line [58]. Such lines are called homogeneous. Applying Ohm's and Kirchhoff's laws, the differential equations for a homogeneous line can be obtained:

$$\frac{\partial V(x,t)}{\partial x} = -r_0 I(x,t) - L\frac{\partial I(x,t)}{\partial t}, \tag{1.1}$$

$$\frac{\partial I(x,t)}{\partial x} = -g_0 I(x,t) - C\frac{\partial I(x,t)}{\partial t}. \tag{1.2}$$

From the differential equations it follows that $V(x,t)$ and $I(x,t)$ are functions containing two variables, so the transmitted signal has a wave nature, and reflections are observed in case of line mismatch. The wave impedance of a homogeneous line is determined by the following formula:

$$Z_0 = \frac{V(x)}{I(x)} = \sqrt{\frac{r_0 + j\omega L}{g_0 + j\omega C}}. \tag{1.3}$$

In modern technologies, the conventional unit of 50 Ohm is chosen as the optimal value of wave resistance in the intermediate range of the smallest losses and the largest transmitted power [59]. The reflection coefficients at the transmitter (K_t) and receiver (K_r) nodes are determined as follows:

$$K_t = \frac{R_{out} - Z_0}{R_{out} + Z_0}, \tag{1.4}$$

1.1 General Issues of Means to Accelerate Transfer of Information...

$$K_r = \frac{R_{input} - Z_0}{R_{input} + Z_0}, \quad (1.5)$$

where R_{out} corresponds to the output resistance of the conducting node and R_{input} corresponds to the input resistance of the receiving node. Therefore, resistance termination blocks (RTBs) are designed, which provide a resistance value of 50 Ω, regardless of supply voltage, temperature, and technological deviations [60–63]. To ensure such accuracy, external resistors placed outside the IC are often used, the accuracy of which varies in the range of $\pm 3\%$.

There are four basic resistance termination architectures [58]:

- Non-terminated architecture
- Termination only in transmitter
- Termination only in the receiver
- Termination both in the transmitter and receiver

In the first case, there is no RTB in either the transmitter or receiver. Such an architecture is designed for low-power systems and works with rather short channels. It allows to save the area on the die due to the absence of RTBs, but it limits the performance of the system. Only the architecture with the RTB implemented in transmitter is intended for medium-performance systems, but in this case, the value of the output voltage of the transmitter is established in two stages, for a period twice the delay time of the channel. When the RTB is implemented only at the receiver, ideally the signal is no longer reflected back to transmitter. However, technological deviations and non-ideality of the system exclude the precise setup of the resistance value, which leads to reflections that do not fade in the transmitter. Two-way built-in RTB ensures minimal reflections and increases system reliability. It is the best architecture to avoid reflections. However, in this case, there is a loss of the signal level twice at the expense of the resulting voltage divider. Such a loss is compensated by a differential signal transmission option. This is the reason why modern high-speed systems use an architecture with a built-in bidirectional RTB and send the signal in a differential form.

The presence of R_0 and g_0 components in a channel is conditional to both thermal and skin effect and dielectric absorption [64, 65]. As a result of the skin effect, the functional resistance of the conductor decreases as the frequency increases:

$$R(f) = R_h * \sqrt{\frac{f}{f_s}}, \quad (1.6)$$

where f_s is the frequency when the predominant part of the current passes through half of the area of the wire and R_h corresponds to the resistance of the wire in case of a constant signal. Therefore, the losses due to the resistive component in coordinated long lines can be represented by the following formula:

$$\alpha_R = \frac{R}{2Z_0} = \frac{R_h}{2Z_0} * \sqrt{\frac{f}{f_s}}. \quad (1.7)$$

It follows from the above formula that the losses due to the skin effect are a function of the frequency and increase along with its increase.

The effect of dielectric absorption is due to the phenomenon of free dipoles absorbing thermal energy. When a direct concentrated electric field is applied, the free dipoles change their direction and absorb thermal energy.

The dielectric absorption effect in aligned long lines can be estimated by the following formula:

$$\alpha_D = \frac{G * Z_0}{2} = \pi * f * C * \tan\delta_D * \sqrt{\frac{L}{C}}. \quad (1.8)$$

As can be seen from the obtained formula, the losses due to dielectric absorption are directly proportional to the frequency.

Calculation of losses due to skin effect and dielectric absorption gives a rather accurate picture to estimate losses of terminated long line (Fig. 1.9).

The obtained graph shows that the signal in long lines is suppressed as the frequency increases. Summarizing, it can be noted that the increase of the operating frequency is limited by the frequency characteristics of long lines. Therefore, there is a need to restore the transmitted signal in the receiver before the reading process. To perform this function, special equalizers are designed in the transmitter and receiver, which will provide equalization of the signal in the operating frequency range.

It is also noteworthy that the four-level Pulse Amplitude Modulation (PAM4) of signal transmission is used in modern sequential information transmission standards [66–72]. It allows to get the same speed, but reduce the frequency of signal transmission twice.

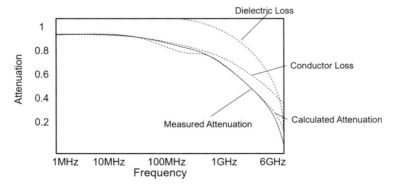

Fig. 1.9 Losses in the long lines

1.1 General Issues of Means to Accelerate Transfer of Information... 9

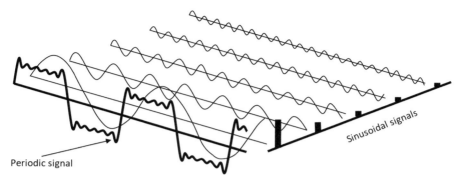

Fig. 1.10 Periodic signal analysis to sinusoidal signals

The Importance of Sequential Information Reception and Equalization in an Integrated Circuit

Every integrable and periodic function can be represented by a Fourier series [73–75].

$$F(t) = \frac{a_0}{2} + \sum_{n=1}^{\infty}(a_n * \cos(n*t) + b_n * \sin(n*t)) = \sum_{n=1}^{\infty} c_n * e^{i*n*t}. \quad (1.9)$$

That is, an arbitrary signal transmitted by I/O cells can be analyzed into an infinite or finite number of sinusoidal signals with different frequencies (Fig. 1.10).

Considering that the frequency characteristics of long lines decrease nonlinearly as the frequency increases, it can be said that different harmonic components of the transmitted signal are modified to different extents.

Therefore, the equalizer, built into the transmitter, should amplify the high-frequency components of the signal and ensure full equalization of the signal in the operating frequency range (Fig. 1.11) [76–78].

Equalization of the fundamental harmonics of the signal is an important condition, because high-frequency components are considered as noise. Therefore, it is necessary to increase the operating frequency and gain of the equalizer in that domain.

To estimate signal distortions, it is convenient to use the eye diagram [79–83]. It represents the superposition of all transmitted bits, which is obtained as a result of dividing the transmitted data by its paragraph size. The advantage of the method is that it allows to estimate the timing parameters of a signal of arbitrary length by means of a graphical representation. Below are the amplitude-frequency characteristics of a 17-inch-long channel and the eye diagram of a signal transmitted at 5 GHz (Fig. 1.12).

The eye diagram shows that the signal is completely distorted and cannot be read at the receiver. In addition to the equalizer, implemented in the receiver node, a digital filter with a finite impulse response of several orders is applied in the

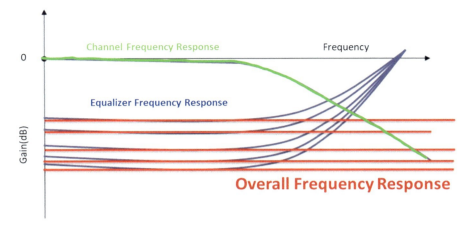

Fig. 1.11 Equalization of signals transmitted in I/O cells

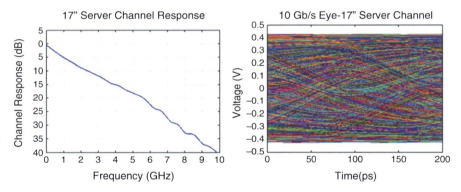

Fig. 1.12 The amplitude-frequency characteristic of the channel (**a**) and the eye diagram of the signal transmitted by it (**b**)

transmitter node, which is able to perform signal equalization to some extent (Fig. 1.13) [84–87].

The presence of a digital filter ensures the dependence of the signal level on the preceding and following bits.

$$W(z) = W_{-1} + W_0 * z^{-1} + W_1 * z^{-2} + \cdots + W_n * z^{n-1}. \quad (1.10)$$

Figure 1.14 shows the eye diagram of the channel, the digital filter implemented in the transmitting node, and the 5 GHz signal.

1.1 General Issues of Means to Accelerate Transfer of Information...

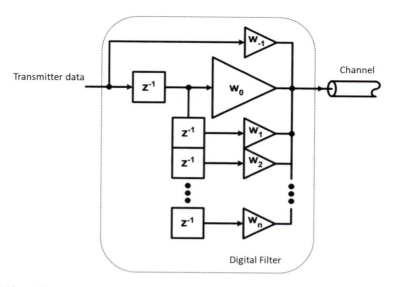

Fig. 1.13 A digital filter with finite impulse response of order n

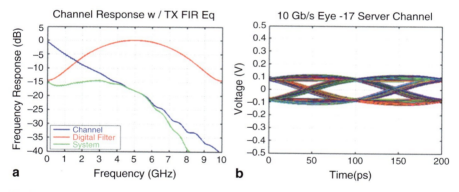

Fig. 1.14 Amplitude-frequency characteristic of channel and digital filter (**a**) and its effect on the signal (**b**)

As it can be seen from the eye diagram, the transmitted signal can be recovered using a sufficiently accurate operating comparator. However, technological deviations and design difficulties do not allow using the mentioned method in more high-performance systems. Therefore, there is an inevitable demand for the development and design of information recovery means at the receiving node. Thus, the design of means of increasing the speed of sequential information reception and processing in the receiving node is one of the problems of modernizing IC development.

1.1.2 Current State and Issues of Existing Means of Increasing the Speed of Receiving Sequential Information in an Integrated Circuit

In systems containing multiple ICs, various measures are used to ensure error-free transmission and reception of data. Their application depends on the standard of signal transmission, as well as on the physical dimensions and amplitude-frequency characteristics of the channel in PCB. Small distortions caused by a sufficiently short channel can be recovered even by means of an analog receiver block (ARB) [88, 89]. Such a choice may also be due to the low frequency of the signal transmitted in the system. Although various low-frequency communication systems still exist today, the increase in speed due to modern signal transmission standards has led to the need to design new methods [90]. However, the technical qualifications of modern standards require that their implementation in the systems also provides the possibility of low-power operation [14–19]. Therefore, the effectiveness of the means depends on the technical conditions of the specific system.

The main parameters of the means of receiving sequential information are the maximum operating frequency and gain, the linearity of the implemented nodes, the area occupied by the IC, its power consumption, and, of course, the cost of the system. It is also necessary to take into account the influence of the noise sources of the node or nodes introduced in the system, which is even more fundamental in the case of the application of four-level amplitude modulation for signal transmission, because in this case the signal voltage reserves are reduced. It is also often necessary to apply several measures together in order to achieve the desired outcome.

Existing Methods to Increase the Speed of Continuous Time Linear Equalizer (CTLE)

The channel has a characteristic of a low-pass filter, and as the frequency increases, the signal distortions become more fundamental. Therefore, there is a need to design a special basic block in the receiver node, which will have a high-pass filter characteristic and will be able to perform signal equalization [91, 92].

In one of the well-known approaches, it is proposed to use active and passive continuous time linear equalizer (CTLE) [93]. A passive CTLE is a combination of high- and low-pass filters made up of resistors and capacitors (Fig. 1.15) [94, 95].

The transfer function of the circuit is:

$$H(s) = \frac{R_2}{R_1 + R_2} * \frac{1 + R_1 * C_1 * s}{1 + \frac{R_1 * R_2}{R_1 + R_2} * (C_1 + C_2) * s}, \quad (1.11)$$

where R_1, R_2, C_1, and C_2 correspond to passive CTLE resistor and capacitor values. It can be seen from the transfer function that it has one zero and one pole, which are determined as follows:

1.1 General Issues of Means to Accelerate Transfer of Information...

Fig. 1.15 Passive CTLE

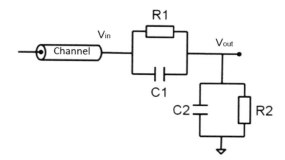

$$\omega_{zero} = \frac{1}{R_1 C_1}, \quad \omega_{pole} = \frac{1}{\frac{R_1 * R_2}{R_1 + R_2} * (C_1 + C_2)}. \quad (1.12)$$

The passive CTLE gain is entirely dependent on resistor and capacitor values.

$$A_s = \frac{R_2}{R_1 + R_2}, A_a = \frac{C_1}{C_1 + C_2}, \quad (1.13)$$

where A_s is the amplification of the system to ω_{zero} frequency and A_a is the gain of the circuit in the range above ω_{pole} frequency. It can be seen from (1.12) that the gain of the passive CTLE at low frequencies depends only on the ratio of resistances and does not affect it in the high frequency range. It is determined by the capacitance ratio. This allows to apply different ratios for the above components and get a positive change in the gain as the frequency increases. To evaluate this change, the ratio between A_a and A_s is considered.

$$\frac{A_a}{A_s} = \frac{\omega_{pole}}{\omega_{zero}} = \frac{R1 + R2}{R2} * \frac{C_1}{C_1 + C_2}. \quad (1.14)$$

However, it follows from (1.13) that the absolute value of the gain is smaller than 1 both at low and high frequencies. Therefore, passive CTLE partially solves the signal equalization problem and cannot provide sufficient gain.

Active CTLE consists of two proportional differential branches connected by a resistor and capacitor (Fig. 1.16) [96]. They are designed to suppress the low-frequency harmonics of the signal.

The transfer function of the circuit is determined by:

$$H(s) = \frac{g_m}{C_{out}} * \frac{s + \frac{1}{R_f C_f}}{\left(s + \frac{1 + g_m R_f / 2}{R_f C_f}\right)\left(s + \frac{1}{R_{load} C_{out}}\right)} \quad (1.15)$$

where g_m is the input transistor conductance, R_{load} is the amplifier load, and R_f and C_f are the feedback resistance and capacitance, respectively.

Fig. 1.16 Active CTLE

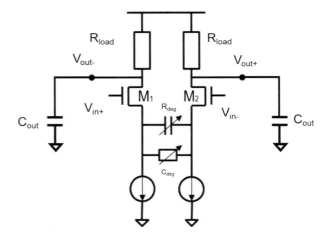

The zero and poles of the transfer function are determined as follows:

$$\omega_{\text{zero}} = \frac{1}{R_f C_f}, \quad \omega_{\text{pole1}} = \frac{1 + g_m R_{\text{deg}}/2}{R_f C_f}, \quad \omega_{\text{pole2}} = \frac{1}{R_{\text{load}} C_{\text{out}}}. \quad (1.16)$$

The formula characterizing the dependence of the maximum value of the circuit gain is determined below:

$$A_a = g_m * R_{\text{load}}. \quad (1.17)$$

At sufficiently low frequencies, when the complex resistance of C_f is sufficiently greater than the resistance of R_f, it can be neglected and the circuit observed as a degraded common-source amplifier. Therefore, the gain of the circuit in the low frequency range is:

$$A_h = \frac{g_m * R_{\text{load}}}{1 + g_m R_f / 2}. \quad (1.18)$$

Obviously, there will be a difference in the gain at low and high frequencies. To estimate this, the relative gain (RG) parameter is introduced, which is determined by the following formula:

$$\text{RG} = \frac{\omega_{\text{pole1}}}{\omega_{\text{zero}}} = 1 + g_m R_f / 2. \quad (1.19)$$

A high value of RG makes it possible to perform signal equalization in the case of longer channels, when the recovery of its high-frequency components becomes more fundamental. But no less important is the possibility of changing the RG, which allows to restore the signal, coming from different channels.

1.1 General Issues of Means to Accelerate Transfer of Information... 15

CTLE RG is controlled by a binary code (Fig. 1.17).

The C_1 capacitor, implemented in the presented structure, provides greater reliability against noises caused by power buses and increases their suppression factor. It allows to filter the relevant harmonics of the noises and as a result keep the gate-source voltage of the M1 transistor constant. Since the M1 transistor is a current source in the differential branch, its constant current value ensures the reduction of the vibration caused by the noise of the power supply jitter of the entire system.

The system also includes a binary code-driven resistance and capacitance matrix (RCM) (Fig. 1.18).

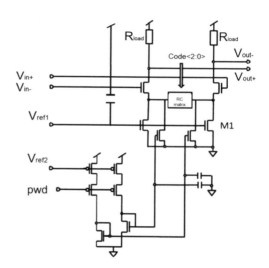

Fig. 1.17 Active CTLE with controlled RG

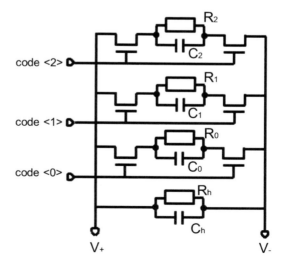

Fig. 1.18 RCM circuit

Fig. 1.19 The block diagram of the sequential information processing algorithm in the receiver

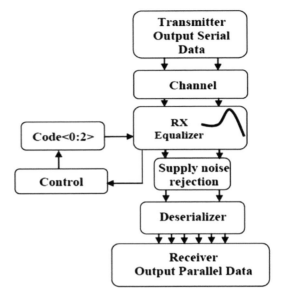

It is a combination of code-controlled high-pass and low-pass filters that allow the overall system gain and therefore the RG to be changed. Thus, the implemented RCM allows to solve the problem of working with different channels. However, this is done due to the reduction of the gain in the low frequency range, which after equalization setup leads to a decrease in the amplitude of the output signal.

Thus, reception and processing of sequential information at the receiver node is performed in phases and by adjusting the RCM control codes (Fig. 1.19).

In the first stage, due to feedback, system setup is performed, during which the optimal codes for signal reception are selected. Then, the noise from the supply voltage is filtered out of the equalized signal, and the sequential information is converted to parallel.

To evaluate the performance of the system, frequency analysis was performed and its amplitude-frequency characteristics were obtained for all possible codes (Fig. 1.20).

It can be seen from Fig. 1.20 that the introduction of the controlled RCM in the system does not affect the maximum gain and allows to control its value at low frequencies. There is also the possibility to control the ω_{zero} frequency, which allows to perform optimal system setup when using different channels.

However, the results are strongly dependent on the technological process, as well as on the change of supply voltage and temperature. All corner cases of CTLE frequency parameters with embedded RCM are summarized in Table 1.1

Simulation of power supply rejection ratio (PSRR) was also performed, the characteristics of which for typical and worst cases are shown in Fig. 1.21.

The obtained results of PSRR indicate that its absolute value is about 35% more than the amount required by the USB standard [14].

1.1 General Issues of Means to Accelerate Transfer of Information... 17

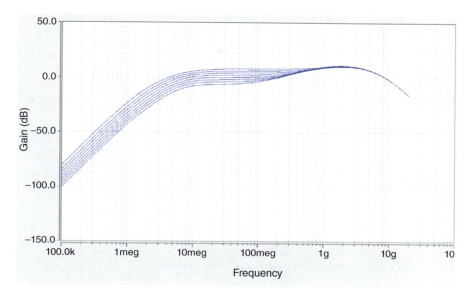

Fig. 1.20 Amplitude-frequency characteristics of the system for all possible codes

Table 1.1 Full amplitude-frequency results of the system

Process/temperature	Maximum RG (dB)	Low-frequency bandwidth (MHz)	High-frequency bandwidth (GHz)
Typical (25 °C)	10.52	9.22	8.38
Slow (125 °C)	8.37	10.35	10.85
Fast (−40 °C)	11.42	10.78	9.57

Thus, the presented method of increasing the speed allows to increase the frequency of the transmitted signal. However, the equalization of the signal using the method ensures a lower level of the output signal, which makes the problem of its processing in the next node more fundamental. Therefore, the applied method cannot completely solve the proposed problems.

Existing Methods to Increase the Speed of the Decision Feedback Equalizer (DFE)

An increase in the frequency of the transmitted signal leads to an increase in the effect of inter-symbol interference (ISI). In the case of ISI, the value of the voltage in each level of the transmitted signal is determined by the superposition of the previous and next bits. Therefore, there is an inevitable requirement to design and implement new nodes for signal recovery and processing. In one of the well-known approaches, it is proposed to introduce a decision feedback equalization circuit (DFE), which is a digital filter with finite impulse response, after the CTLE in the receiver node [97, 98].

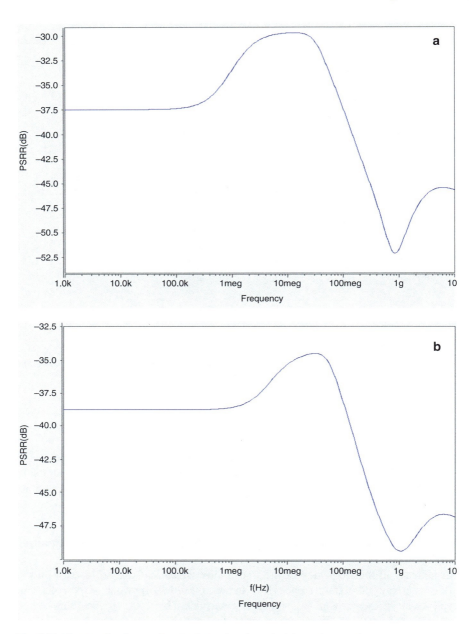

Fig. 1.21 Suppression factor of power bus noises in typical (**a**) and worst (**b**) cases

Figure 1.22 shows the architecture of the receiver node with the embedded *DFE*. The code-controlled RTB provides the required channel termination and common mode level of the input signal. CTLE then performs signal equalization for the given

1.1 General Issues of Means to Accelerate Transfer of Information...

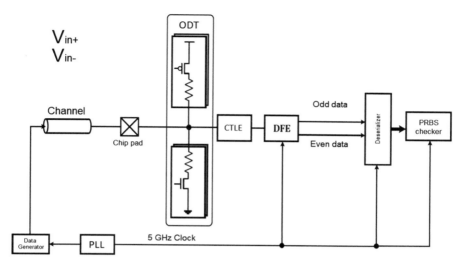

Fig. 1.22 Receiver node architecture with embedded DFE

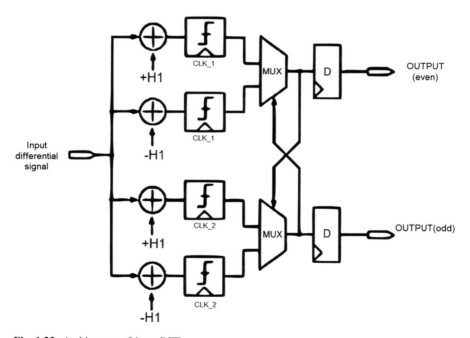

Fig. 1.23 Architecture of 1-tap DFE

channel. And the 1-tap DFE filters the noise generated by the ICI from the equalized signal and restores the fully differential signal.

The 1-tap DFE architecture is shown in Fig. 1.23.

For even and odd taps of transmitted information, two branches are implemented, the selection of which is made using multiplexers. Since feedback timing parameters are a primary concern in DFEs implemented in modern high-speed I/O cells, it was decided to use the output of the multiplexers as the control signal of the parallel branch [99]. As a result, the formed feedback coefficient (H1) is added (or subtracted) to the input signal, and the decision is made already taking into account the digital value of the previous bit.

Figure 1.24 shows the architecture of comparators implemented in a 1-tap DFE.

The presence of N1a and N1b transistors provides isolation between input and output signals. And N3a and N3b transistors isolate the input pair from the clock signal. Thus, the above-mentioned transistors increase the noise resistance of the system.

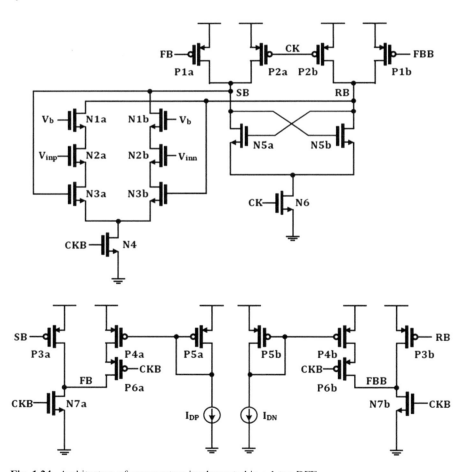

Fig. 1.24 Architecture of comparators implemented in a 1-tap DFE

1.1 General Issues of Means to Accelerate Transfer of Information... 21

The I_{dp} and I_{dn} reference currents are controlled by digital-to-analog converters (DACs) (Fig. 1.25).

Reference current control provides capability to adjust the difference between the output voltages caused by the non-ideality of the manufacturing process. The experimental results prove that the current DAC can compensate the deviation voltage up to 50 mV.

Transient simulation was performed to evaluate the performance of the system, and the obtained results were summarized in eye diagrams (Fig. 1.26).

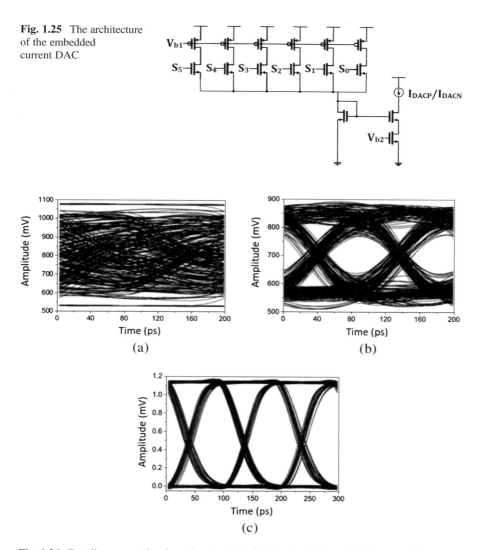

Fig. 1.25 The architecture of the embedded current DAC

Fig. 1.26 Eye diagrams at the channel output (**a**), CTLE output (**b**), and DFE output (**c**)

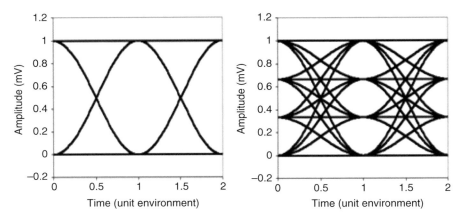

Fig. 1.27 2 (**a**) and 4 (**b**) PAMs

The simulation results prove that the architecture of the proposed receiver, which consists of a CTLE and a 1-tap DFE, allows to realize sequential information reception at a frequency of 5 GHz.

Thus, the implementation of a 1-tap DFE in the receiver allows to compensate the noise caused by ICI and to increase the transmission frequency of the signal. However, even with the use of multi-tap DFE, the speed of signal transmission is limited by the speed of the elements in the system. Therefore, the applied method has clear limitations, and there is a problem of designing an architecture with elements with higher speed.

Existing Methods to Increase the Speed of Reception of Sequential Information Using Pulse Amplitude Modulation (PAM4)

PAM2 has been used for signal transmission for decades. However, the amplitude-frequency characteristics of the channel show that the increase in the frequency of the signal is limited by practically irreversible distortions. For this reason, over time, they also started to use PAM4, which allows to transmit twice as much information at the same signal frequency [100, 101] (Fig. 1.27).

In the case of PAM4, each signal level corresponds to 2 bits of information. Therefore, the same data transmission speed can be obtained by using twice lower frequency. However, such an approach leads to a reduction of the voltage reserve between two neighboring levels, which further tightens the noise immunity requirements.

Figure 1.28 shows the architecture of a PAM4-enabled receiver.

In the first stage, equalization of distorted signal is implemented using the CTLE. Then, the embedded three comparators, which have different offsets, perform a signal level check. The checked values are stored in subsequent triggers, which are controlled by the synchronizing signal generated by the embedded PLL. The values

1.1 General Issues of Means to Accelerate Transfer of Information... 23

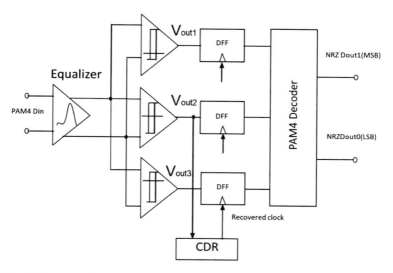

Fig. 1.28 Architecture of a PAM4-enabled receiver

Fig. 1.29 VSA architecture

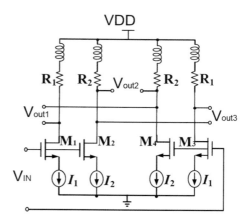

stored in the triggers are transformed from thermometric to unary code by means of a decoder.

In order to perform a three-stage comparison, a voltage shifting amplifier (VSA) and an output buffer were used. M1/M2 and M3/M4 transistors embedded in VSA are not differential pairs, because the reference currents in their branches (I1~I4) are different (Fig. 1.29). Such a structure amplifies the signal with the appropriate amplitude and provides only its variable component so that the output buffer makes the right decision.

Figure 1.30 shows the results of transient simulation.

The complete system results are presented in Table 1.2.

Fig. 1.30 Eye diagrams in CTLE output (**a**), comparator output (**b**), and decoder output (**c**)

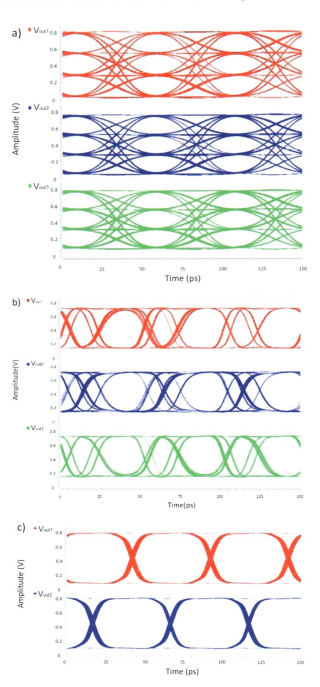

1.1 General Issues of Means to Accelerate Transfer of Information... 25

Table 1.2 Complete results of the receiver

Parameter	Results obtained by implementing the existing method
Operating frequency (Gbps)	38.4~40.4
Jitter of the clock signal in the receiver (ps, p-p)	2.3 (at 20 GHz)
Jitter of the information signal in the receiver (ps, p-p)	3.2 (at 20 Gbps)
Surface area occupied on a die (mm^2)	1 × 0.7
Technology	SAED 32 nm

Thus, the method of using the PAM4 of the signal makes it possible to significantly increase the speed of data transfer. The architecture of the presented receiver allows to perform PAM4 signal reception, processing, and transformation into a binary code. However, when using the method, there is a need to design a high-precision and high-speed three-level comparator, as the voltage margin between adjacent signal levels decreases. In the case of such an architecture, the speed of the receiving node is also limited by the use of nodes implemented in the comparator.

1.1.3 Proposed Approaches to Increase the Speed of Receiving Sequential Information in an Integrated Circuit

The existing methods and approaches to increase the speed of receiving sequential information in an integrated circuit do not fully meet its current requirements. Therefore, the development of new solutions and methods continues to be a current issue. Based on this, the following approaches have been proposed:

1. Implementation of a neg-C circuit, controlled by a current DAC in CTLE (Fig. 1.31).

Such an architecture will ensure a decrease in the output complex resistance of the CTLE and, therefore, an increase in the bandwidth of the entire system. This, in turn, will increase the gain of the system at operating frequencies. In order to ensure the same gain in the entire operating range of the output signal, there will be an inevitable need to adjust the linearity of the implemented circuits. Adjustment of the thermometric branches in the current DAC will ensure the linearity of the output current and exclude the loss of the code. And in *neg-C circuit*, the linearity setup system controlled by the constant component of the signal will ensure the uniform change of its currents in the differential branches. As a result, with an increase in the area occupied on the die and the power consumption within the permissible limits, such an architecture will allow to equalize the signal with a higher frequency and a larger output operating range.

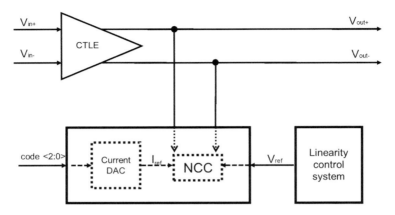

Fig. 1.31 Architecture of CTLE with embedded neg-C circuit

Fig. 1.32 Architecture of the proposed DFE

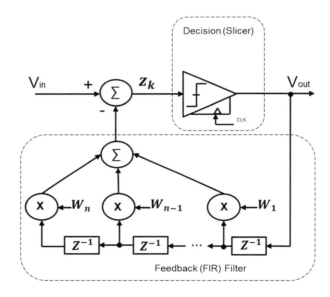

2. Building DFE architecture with specially designed high-speed comparators (Fig. 1.32).

The proposed comparator architecture will provide low input capacitance and faster output signal switching. After switching, the voltage difference caused by charge diffusion of different levels of output signals and in the channels of transistors that function as a switch will be adjusted by a special system controlled by the clock signal. It will also allow to reduce the offset in the system. Thus, the proposed DFE architecture, due to the nonsignificant increase in power consumption, will reduce the timing constraints of the entire system and allow to increase the frequency of the received signal.

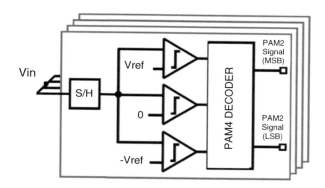

Fig. 1.33 Architecture of a receiver with parallel branches

3. Application of higher-speed receiver architecture, operating with PAM4 (Fig. 1.33).

In order to reduce the timing constraints during the process of determining the order of the equalized signal at the receiver, it is recommended to use four proportional branches that will work in sequence. The sample-hold blocks in them, which work with a synchronous signal with a uniform phase shift, will record the level of the signal in the given branch until the end of the given order decision. Therefore, in the case of using such an architecture, due to the increase in the area occupied on a die within the permissible limits, it will be possible to reduce the timing constraints of the system by four times.

The above-mentioned approaches are aimed at increasing the speed of data transfer. They will allow to perform equalization of the signal with a higher frequency and reduce the time limitations of data processing.

Conclusions

1. The design of means for restoring the distorted signal has become extremely relevant at present, because due to the increase in the amount of data processed in high-speed systems, in order to meet the demand for increasing the speed of their transmission, the speed of reception of sequential information is limited by the amplitude-frequency characteristics of the channel.
2. Experimental studies of the existing means of increasing the speed of receiving sequential information and the results of their analysis show that they do not completely solve the problems in integrated circuits and have a number of time limitations that are unacceptable from the point of view of practical application. That circumstance dictates the need to develop more effective means of the mentioned class.
3. Effective approaches from the point of view of time limitations for the development of means of increasing the speed of sequential information reception in integrated circuits have been proposed. The embedded nodes and the structure of their architecture provide an opportunity to significantly increase data transfer and processing frequencies in the case of increasing the area occupied on the die and power consumption within the permissible limits.

1.2 Methods to Increase the Speed of Receiving Sequential Information in an Integrated Circuit

1.2.1 Method to Increase the Speed of Receiving Sequential Information with the Use of a Circuit with Negative Capacitance in Continuous Time Linear Equalizer (CTLE)

The discussions of the means to increase the speed of CTLE and the research on its application prove that it has clear limitations and cannot fully meet the modern standards of data transfer. Built-in RCM enabled higher-frequency signal equalization, reducing the output operating range.

In order to implement equalization of the transmitted signal with a higher frequency, it has been proposed to design and include at the output of the CTLE a binary code-controlled negative capacitance circuit (neg-C circuit). The simplest neg-C circuit consists of a capacitor and two transistors connected with positive feedback (Fig. 1.34) [102].

The positive feedback in the system allows to obtain a negative value of the complex resistance. The impedance of a neg-C circuit is:

$$Z_{eq} = -\frac{1}{sC} \frac{g_m + s(C_{gate-source} + 2C)}{g_m - sC_{gate-source}}, \quad (1.20)$$

where g_m is the conductance of the feedback transistor and $C_{gate-saource}$ is the gate-source parasitic capacitance of that transistor.

In order to control the amplitude-frequency characteristics of the *neg-C circuit*, as well as to exclude the influence of the common mode component of the current on the CTLE, it is proposed to apply the following architecture (Fig. 1.35) [103].

To make the reference current controllable with a digital code, it is recommended to use an 8-bit current DAC (Fig. 1.36). The choice of its range is determined by the saturation margin of the current sources in the *neg-C circuit*.

Fig. 1.34 Simplest neg-C circuit

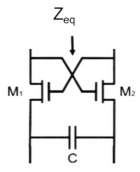

1.2 Methods to Increase the Speed of Receiving Sequential Information... 29

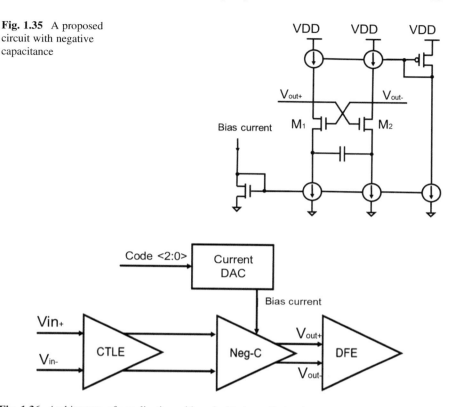

Fig. 1.35 A proposed circuit with negative capacitance

Fig. 1.36 Architecture of equalization with embedded neg-C circuit

As the reference current changes, the g_m of M1 and M2 transistors increases, which leads to amplification of the effect of the formed positive feedback and, therefore, to the increase in the absolute value of the complex resistance of the neg-C circuit. This allows to increase the maximum value of the CTLE gain and the bandwidth.

For a typical case, the amplitude-frequency characteristics of the system and the values of the main parameters in the case of all codes of ADC control are presented in Fig. 1.37 and Table 1.3.

The amplitude-frequency characteristics of the system were also measured in the slow and fast cases (Figs. 1.38 and 1.39).

The obtained results show that in the case of a typical corner, the implemented *neg-C circuit* allows to increase the gain of the system to approximately 10 dB for a signal with a frequency of 5 GHz.

In order to get a complete picture of the system, its amplitude-frequency characteristics were also considered in the case of all possible configuration options of RCM (Fig. 1.40).

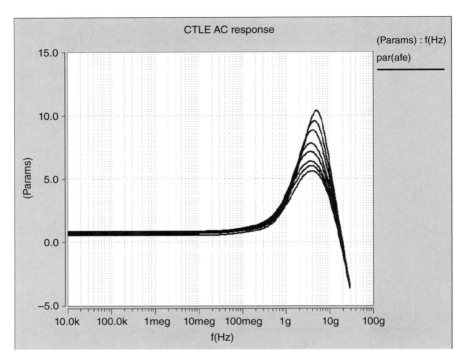

Fig. 1.37 Amplitude-frequency characteristics of the proposed system

Table 1.3 Basic system parameters

DAC control code value	Maximum gain frequency (GHz)	Gain (dB) at 4 GHz	Gain (dB) at 5 GHz
0	3.63	5.6	5.46
1	3.63	5.98	5.8
2	3.63	6.35	6.13
3	3.8	7.1	6.81
4	3.98	7.75	7.44
5	4.17	8.79	8.56
6	4.57	9.45	9.45
7	5.01	9.88	10.36

As a result of frequency analysis, measurements show that as a result of a complete setup of the system, its RG can be increased up to ~15 dB. There is also a possibility to increase its bandwidth up to ~5 GHz.

The parameters describing the performance of the system were measured in the worst corner cases (Table 1.4).

In order to evaluate the noise immunity of the overall system, noise simulation was performed, and in the worst corner case, its mean square value at the CTLE output was measured (Fig. 1.41).

1.2 Methods to Increase the Speed of Receiving Sequential Information... 31

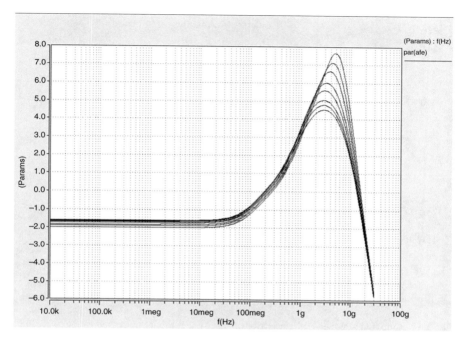

Fig. 1.38 Amplitude-frequency characteristics of the system in the slow case

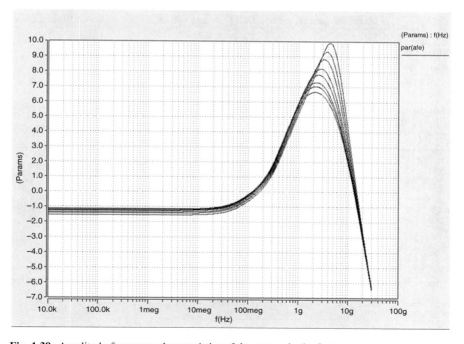

Fig. 1.39 Amplitude-frequency characteristics of the system in the fast case

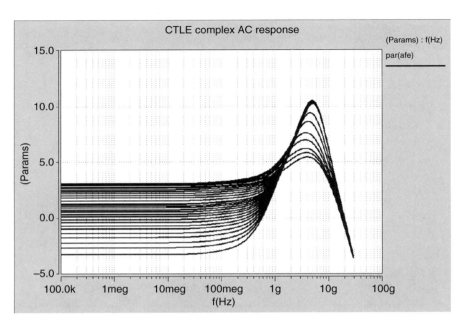

Fig. 1.40 Amplitude-frequency characteristics of the system using the method

Table 1.4 Basic parameters of the system in worst-case corner cases

PVT	Maximum gain frequency (GHz)	Gain (dB) at 4 GHz	Gain (dB) at 5 GHz
Typical, 25	5.01	9.88	10.36
Slow, 125	4.37	7.57	7.59
Fast, −40	6.13	12.2	11.98

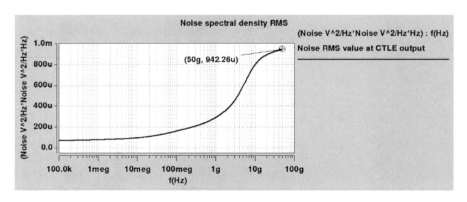

Fig. 1.41 Amplitude-frequency characteristics of the system using the method

1.2 Methods to Increase the Speed of Receiving Sequential Information... 33

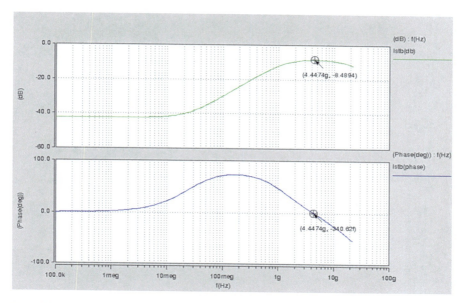

Fig. 1.42 Stability characteristics of the system in a typical case

In order to evaluate the stability of the *neg-C circuit*, frequency analysis was performed (Figs. 1.42 and 1.43).

However, all the above results were obtained in the case of small signal simulation, so it is necessary to evaluate the linearity of the system.

The linearity of amplifiers is measured with a 1 dB compression point (Fig. 1.44) [104].

The 1 dB compression point is the input signal amplitude level in the result of which the gain is 1 dB less than the nominal value. However, the level of the output signal at that point is also important. It must be high enough to include the input operating range of the next node. Since the CTLE is followed by a decision feedback equalizer (DFE) and operates with an input signal amplitude of up to 150 mV, the output signal amplitude should be no less than 150 mV for a 1 dB compression point [68].

The M1 and M2 transistors included in the neg-C circuit have the same dimensions and provide the same current in the differential branches at the same voltage drop. In the presence of a variable component of the input voltage, the voltages controlling them change, which allows for faster switching due to positive feedback. However, at low values of the voltage, the M1 and M2 transistors appear in cut-off mode, and the amount of deviation with respect to the constant component between the differential branches is violated.

It is recommended to connect M3 and M4 transistors parallel to M1 and M2 transistors, which will be controlled only by the constant component of the input signal (Fig. 1.45).

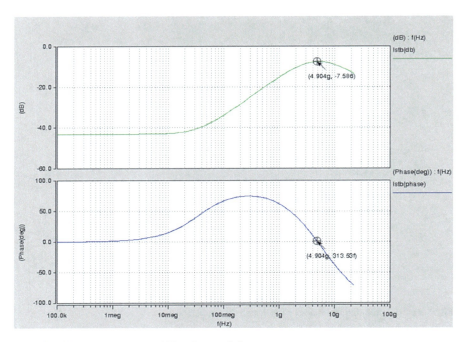

Fig. 1.43 Worst-case system stability characteristics

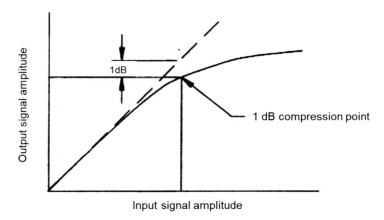

Fig. 1.44 1 dB compression point

During the simulation, a sinusoidal signal was given to the input of the CTLE, the amplitude of which starts to increase over time (Fig. 1.46).

As it can be seen from the graph of currents of M1 and M2 transistors, they increase and decrease unevenly (Fig. 1.47).

Fig. 1.45 The linearity setup system included in the neg-C circuit

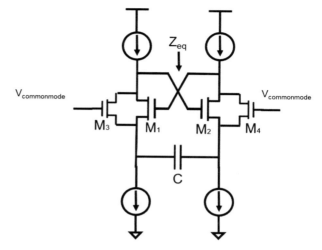

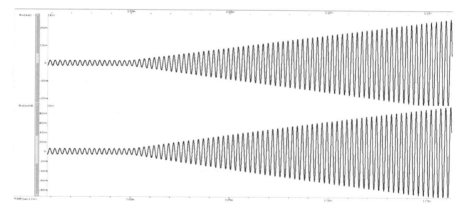

Fig. 1.46 Input and output signals of the system

The presence of M3 and M4 transistors weakens the effect of positive feedback, but ensures the stability of the current of differential branches. As a result, the currents of the differential branches increase and decrease in the same proportion, which ensures the linearity of the entire system.

1 dB compression point measurements were made before and after using the method (Figs. 1.48 and 1.49).

In order to evaluate the change of the gain of the system, frequency analysis was also performed, the results of which are summarized in Table 1.5.

The linearity of the system also depends on the implemented current DAC. In the case of an ideal linear current DAC, each increase in the order of the controlling digital code corresponds to an increase in the output current by the corresponding amount of LSB – last significant bit (LSB) [105]. In general, for an N-bit current DAC, the LSB is determined by:

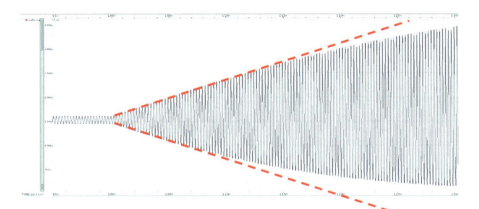

Fig. 1.47 Currents of M1 and M2 transistors

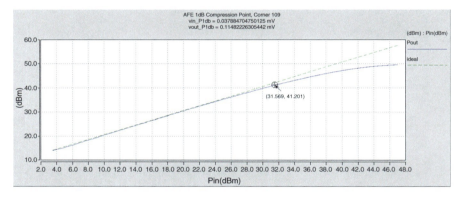

Fig. 1.48 1 dB compression point measurements before applying the method

$$\text{LSB} = \frac{I_{2*N-1} - I_0}{N^2}. \quad (1.21)$$

The main parameters characterizing the linearity of the current DAC are the differential (DNL) and integral (INL) nonlinearities.

$$\text{DNL} = \frac{I_{i+1} - I_i}{\text{LSB}} - 1, \quad (1.22)$$

$$\text{INL} = \frac{I_i - I_0}{\text{LSB}} - i. \quad (1.23)$$

The embedded current DAC architecture is composed of binary and thermometric branches (Fig. 1.50).

1.2 Methods to Increase the Speed of Receiving Sequential Information...

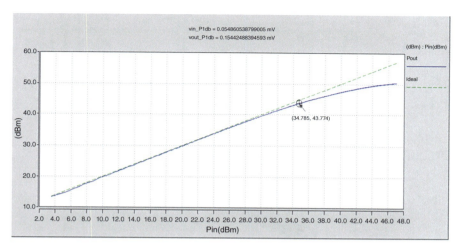

Fig. 1.49 1 dB compression point measurements after applying the method

Table 1.5 Simulation results of linearity setup system implementation

Parameter	Without using the proposed method	Using the proposed method	Difference
Gain (dB) at 4 GHz	9.88	9.25	−0.63 (−6.3%)
Gain (dB) at 5 GHz	10.36	9.8	−0.56 (−5.4%)
Output signal level at 1 dB compression point (mV)	114.8	154.4	39.6 (+34.5%)

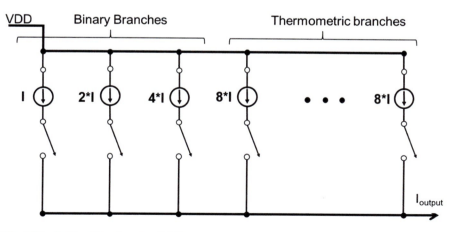

Fig. 1.50 Architecture of current DAC

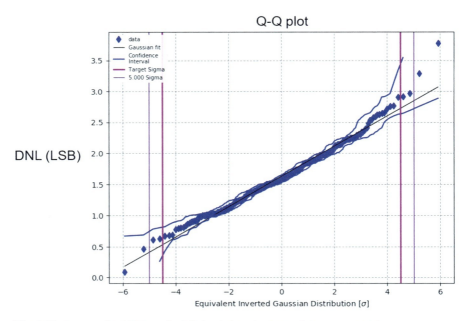

Fig. 1.51 As a result of "Monte Carlo" simulation, the DNL of the current DAC

CMOS transistors operating in the saturation mode are used as a current source. Branch currents are determined by selecting transistor sizes.

Since the nominal currents of the thermometric branches are eight times larger than the size of the LSB, the inaccuracies caused by the technological process in them significantly reduce the nonlinearity of the system.

As a result of "Monte Carlo" simulation, the DNL of the current DAC was measured (Fig. 1.51).

The obtained results indicate that there is a possibility to have DNL up to three LSB. It is an unacceptable value, because in that case the current sources included in the neg-C circuit deviate from the operating mode. It is recommended to reduce the currents of the thermometric branches and obtain a negative DNL when switching them (Fig. 1.52) [106].

The obtained results show that the maximum value of DNL after applying the method is less than one LSB (Fig. 1.53).

Thus, a CTLE setup system providing signal equalization was proposed. The proposed system is based on the reference current control of neg-C circuit including positive feedback, which is implemented by the current DAC. The ability to adjust the system with a digital code provides signal equalization for different channels and data transmission frequencies. Neg-C circuit and current DAC linearity setup methods are proposed, which ensure the saturation condition of the current source

1.2 Methods to Increase the Speed of Receiving Sequential Information... 39

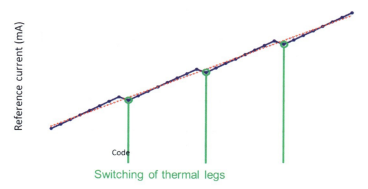

Fig. 1.52 DNL setup method of current DAC

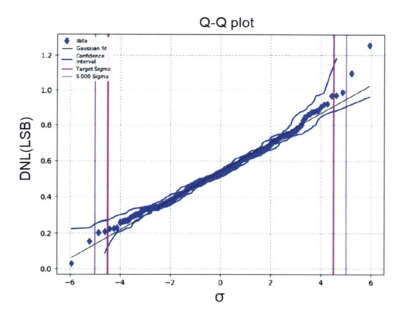

Fig. 1.53 DNL of current DAC as a result of Monte Carlo simulation after application of the method

transistors and, therefore, the increase in the output operating range of the overall system. The proposed system provides approximately two times faster signal equalization. When using the proposed method, compared to existing solutions, its power consumption increases slightly, approximately 10%.

1.2.2 Method to Increase the Speed of Receiving Sequential Information with the Use of a High-Speed Comparator with Low Input Capacitance in the Decision Feedback Equalizer (DFE)

The performance of DFE is limited by timing parameters of cells in it [68]:

$$t_{cl2out,c} + t_{deltime,mul} + t_{stime,t} < 2 \text{ UTI}, \tag{1.24}$$

$$t_{cl2out,t} + t_{deltime,adder} + t_{stime,c} < 1 \text{ UTI}, \tag{1.25}$$

where $t_{cl2out,c}$ is the delay from the comparator clock signal to the output, $t_{deltime,mul}$ is the multiplexer delay time, $t_{stime,t}$ is the trigger setup time, $t_{cl2out,t}$ is the delay from the trigger clock signal to the output, $t_{deltime,adder}$ *is the adder delay time*, $t_{stime,c}$ is the trigger setup time, and UTI is the unit time interval. To reduce $t_{cl2out,c}$, it was proposed to use a circuit of a high-speed comparator with low input capacitance (Fig. 1.54) [107].

In the recovery phase, the np and nn wires connect to the supply voltage through the Mr1 and Mr2 transistors and close the M9 and M10 transistors. At the same time, M11 and M12 transistors reset the potentials of fp and fn wires. Therefore, a value corresponding to the supply voltage is formed at the outputs. This allows reducing the dependence of the comparator on the supply voltage at this stage.

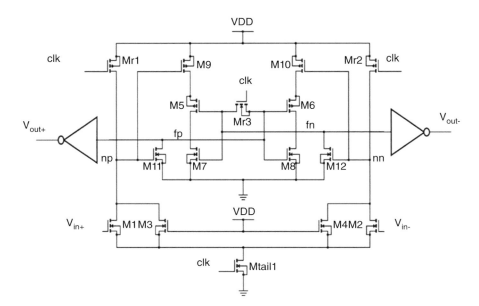

Fig. 1.54 Proposed comparator circuit

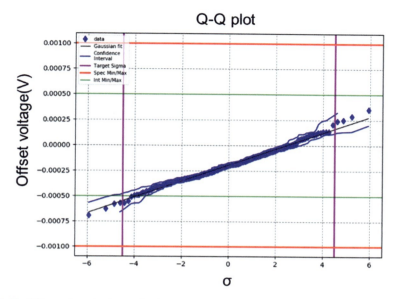

Fig. 1.55 Offset voltage in the result of Monte Carlo simulation for a typical case

In the comparison stage, the M1 and M2 transistors start discharging the np and nn wires. The difference in the input signal provides different current values in the differential branches. M5, M6, M7, and M8 transistors are used to form the output signal.

The constant current induced by M3 and M4 transistors provides faster switching of the output signal.

Comparator offset was measured in the result of Monte Carlo simulation (Figs. 1.55 and 1.56).

Timing simulations were performed on the existing and proposed comparators (Fig. 1.57).

The obtained results indicate that the delay time of the comparator has decreased by about 36%. However, due to the transistors added to it, the total system area on the die has increased by approximately 11%.

Timing simulation of the proposed comparator with changes in technological process, temperature, and supply voltage has been carried out (Fig. 1.58).

A receiver architecture composed of a dual-cascade CTLE and a DFE is proposed [104, 105]. Timing simulation of the whole system was performed (Figs. 1.59 and 1.60).

In order to evaluate the noise immunity of the system, "Monte Carlo" simulation was performed (Figs. 1.61 and 1.62).

As a result of the implementation of the proposed comparators, the layout of the DFE increased by about 8% (Fig. 1.63) [104, 105].

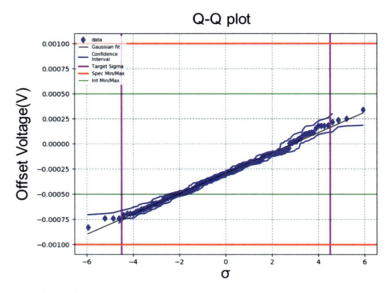

Fig. 1.56 In the result of Monte Carlo simulation, the offset voltage in the worst case

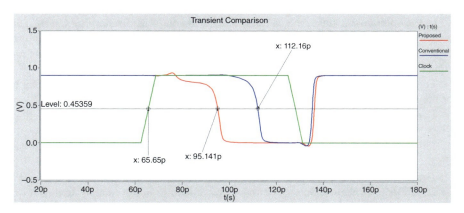

Fig. 1.57 Comparison of delay times of existing and proposed comparators

Thus, an architecture consisting of a two-stage CTLE and a 1-tap DFE providing signal equalization was developed. Its operation is based on low-input high-speed comparators that operate on odd and even data branches in parallel. The digital value of the previous bit determines the amount of influence of the adder in a given branch, as well as the selection of the correct branch. It allows to reduce the ICI caused by the channel and perform correct data processing.

1.2 Methods to Increase the Speed of Receiving Sequential Information... 43

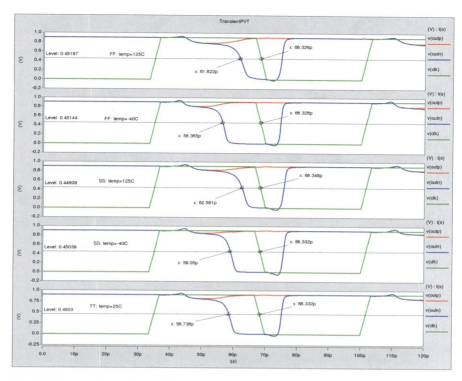

Fig. 1.58 Proposed comparator delay times for extreme corner cases

Fig. 1.59 Eye diagram in the CTLE output

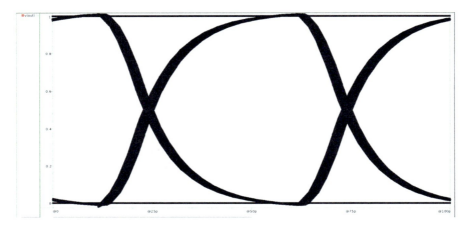

Fig. 1.60 Eye diagram at the DFE output

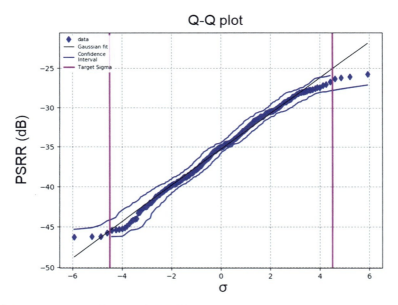

Fig. 1.61 Power supply rejection ratio (PSRR) in a typical case

The use of the proposed comparators reduced the timing constraints in DFE by about 35%. As a result, at the expense of a slight increase in the area occupied on a die and an almost 16% increase in power consumption, twice-higher-speed system was obtained.

1.2 Methods to Increase the Speed of Receiving Sequential Information... 45

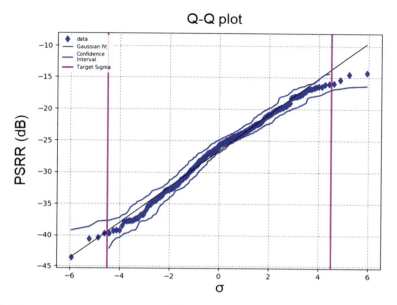

Fig. 1.62 Power supply rejection ratio (PSRR) in the worst case

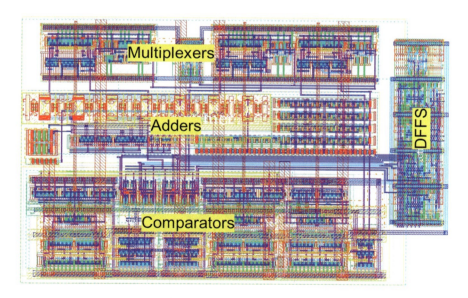

Fig. 1.63 Layout of the proposed DFE

1.2.3 Method to Increase the Speed of Receiving Sequential Information Due to Processing of the Signal Transmitted by Pulse Amplitude Modulation (PAM4), Conditioned by Parallel Branches

As discussed in Sect. 1.1, PAM4 signal transmission enables to increase the amount of received data by two times at the same frequency. It is recommended to apply four identical branches in parallel after the CTLE in the receiver, in which the value of the signal will be captured by means of sample-hold (S/H) blocks (Fig. 1.64).

Instead of the previously considered three-state comparator, a system consisting of three comparators was used, the negative inputs of which are provided with reference voltages [108]. The outputs of the system are transformed into a binary code by means of a decoder. As a result, sequential signal processing and parallel representation transformation are performed.

In comparators, in order to exclude wrong reading of the signal due to deviations of the duty cycle of clock signal, it was adjusted using duty cycle regulator (DCR) (Fig. 1.65) [109].

Due to the negative feedback in the system, the duty cycle detector (DCD) converts the offsets of the output signals from their average value into a control voltage, which is applied to the input of the regulator. Due to this, the constant voltage components of the input signals are equalized. Then the output signal duty cycle is setup. The received incomplete differential signal is converted into a fully differential signal by means of the CML-CMOS buffer.

The constant component of the input voltage in DCR is determined by means of a voltage divider consisting of resistors (Fig. 1.66). The input capacitance provides filtering of the constant component of the signal.

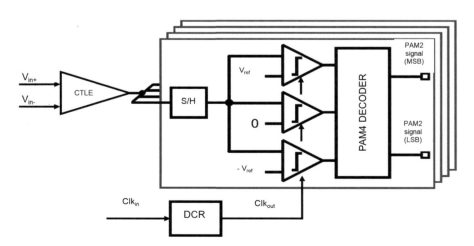

Fig. 1.64 The architecture of the proposed receiver

1.2 Methods to Increase the Speed of Receiving Sequential Information... 47

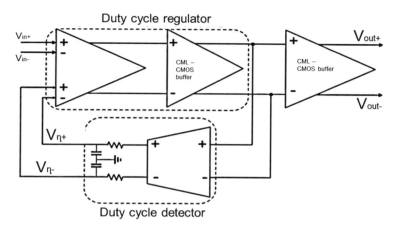

Fig. 1.65 Architecture of implemented DCR

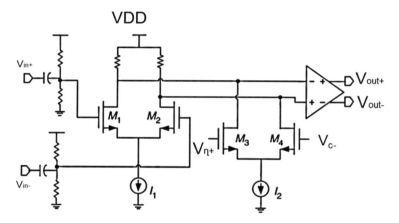

Fig. 1.66 DCR architecture

A special self-adjustment method (Fig. 1.67) [110] was introduced in order to setup the deviation of rise/fall of the output signals of the CML-CMOS buffers due to process deviations.

DCD is a dual-cascade differential amplifier system (Fig. 1.68).

In order to evaluate the speed and accuracy of the system, timing simulation was performed (Figs. 1.69 and 1.70).

After DCR setup, the duty cycle error of the clock signal is less than 0.05% (Fig. 1.71).

In order to evaluate the noise immunity of the system, frequency analysis was performed (Figs. 1.72 and 1.73).

Timing simulation of the entire system was performed (Figs. 1.74 and 1.75).

The complete results of the system are summarized in tabular form (Table 1.6).

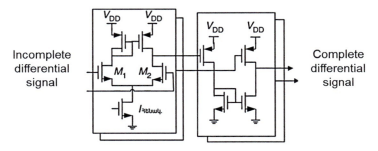

Fig. 1.67 CML-CMOS buffer architecture

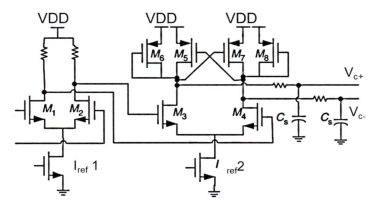

Fig. 1.68 DCD architecture

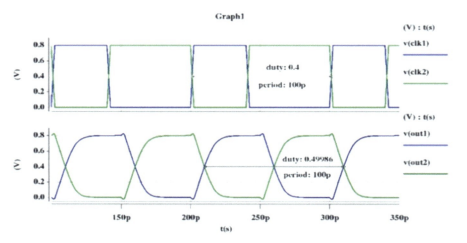

Fig. 1.69 Input and output signals having 0.4 duty cycle

1.2 Methods to Increase the Speed of Receiving Sequential Information...

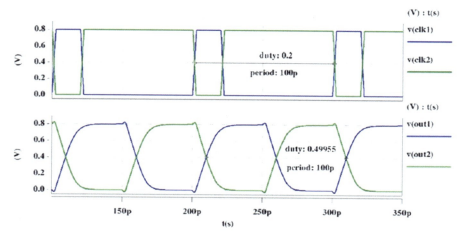

Fig. 1.70 Input and output signals having 0.2 duty cycle

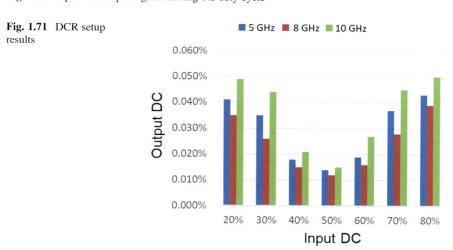

Fig. 1.71 DCR setup results

Thus, a receiver architecture providing equalization and processing of the signal transmitted by PAM4 was developed. Including four identical and parallel operating branches, it allows to perform signal modification and present it in a binary code more quickly. Correction of duty cycle of the clock signal for the control of comparators has been made to eliminate the erroneous reading of the distorted signal due to it. In the DCR output buffers, a self-adjusting system for correcting signal rise/falls has been implemented. High-speed comparators with low input capacitance were also used for signal processing.

The application of the proposed multi-branch architecture, compared to the previous method, made it possible to increase the speed of data transfer by about 50%. The area of the receiver on a die increased by only ~12.2%.

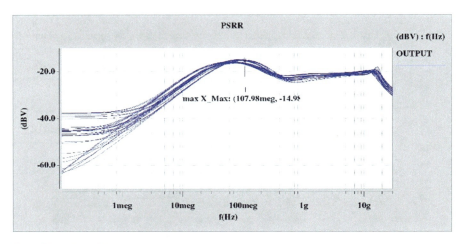

Fig. 1.72 DCR PSRR

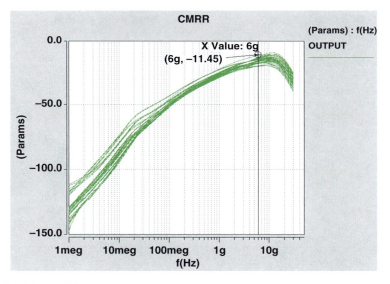

Fig. 1.73 DCR CMRR

Conclusions

1. Effective approaches from the point of view of time limitations for the development of means to increase the speed of sequential information reception in integrated circuits have been proposed. The embedded nodes and architecture structure underlying them provide an opportunity to significantly increase data transfer and processing frequencies in the event of an increase in the area occupied on the die and power consumption within the permissible limits.

1.2 Methods to Increase the Speed of Receiving Sequential Information...

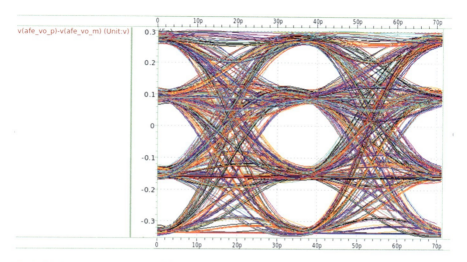

Fig. 1.74 Eye diagram in the CTLE output

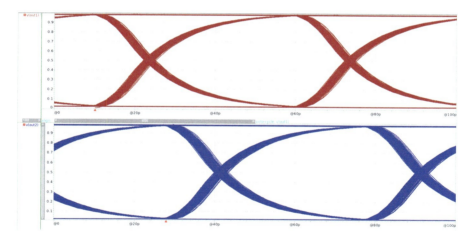

Fig. 1.75 Eye diagram at the decoder output

Table 1.6 Full results for the proposed receiver

Parameter	Results obtained by implementing the existing method
Operating frequency (Gbps)	57.1~60
Area occupied on a die (mm^2)	0.97 × 0.81
Technology	SAED 14 nm

2. A binary code-controlled neg-C implemented system in the continuous time linear equalizer (CTLE) was proposed, which, due to positive feedback and linearity setup, provides approximately two times faster signal equalization at the expense of only 11% increase in the area occupied on the semiconductor crystal.
3. A method to increase the speed of decision feedback equalizer (DFE) was proposed, which, due to the introduction of high-speed comparators with low input capacitance, increases the frequency of data processing by about two times, with only a 16% increase in power consumption.
4. A method of constructing a Pulse Amplitude Modulation (PAM4) transmitted signal processing system has been developed, which, due to data storage and parallel branch processing, increases the data transfer rate by approximately 50% with an approximately 12.2% increase in the area of the receiver on a semiconductor crystal.

References

1. V. Stojanović, *High-Speed Serial Links: Speed Serial Links: Design Trends and Challenges* (Massachusetts Institute of Technology, Integrated Systems Group, 2005), pp. 38–42
2. W. Zhikai, W. Hu, Z. Gu et al., Large-scale integrated circuits simulation based on CNT-FET model. 2021 International conference on IC design and technology (ICICDT) (2021), pp. 1–4
3. B. Zhang, Z. Wentong, Z. Le, et al., Review of technologies for high-voltage integrated circuits. Tsinghua Sci. Technol. **27**, 495–511 (2021)
4. Y. Liang, S. Ruize, Y. Yee-Chia et al., Development of GaN monolithic integrated circuits for power conversion. 2019 IEEE custom integrated circuits conference (CICC) (2019), pp. 1–4
5. Y. Honda, G. Masahide, W. Toshihisa et al., Triple-stacked silicon-on-insulator integrated circuits using Au/SiO$_2$ hybrid bonding. 2019 IEEE SOI-3D-subthreshold microelectronics technology unified conference (S3S) (2019), pp. 1–3
6. M. Rashdan, F. El-Sayed, M. Salman, Performance comparison between SerDes and time-based serial links. 2020 7th international conference on electrical and electronics engineering (ICEEE) (2020), pp. 37–41
7. N. Zhou, K. Huang, F. Lve et al., A 76 mw 40-GB/s SerDes transmitter with 64: 1 mux in 65-nm Cmos technology. 2016 6th international conference on electronics information and emergency communication (ICEIEC) (2016), pp. 155–158
8. A.R. Chada, B. Mutnury, D. Wallace et al., Simulation challenges in designing high speed serial links. 2012 IEEE 62nd electronic components and technology conference (ECTC) (2012), pp. 153–159
9. K. Huang, D. Luo, Z. Wang et al., A 190 mW 40 Gbps SerDes transmitter and receiver chipset in 65 nm CMOS technology. 2015 IEEE custom integrated circuits conference (CICC) (2015), pp. 1–4
10. A.A.S. SH, K.S. Reddy, A 20 Gb/s latency optimized SerDes transmitter for data centre applications. 2020 IEEE international conference on electronics, computing and communication technologies (CONECCT) (2020), pp. 1–4
11. A. Bandiziol, W. Grollitsch, F. Brandonisio et al., Design of a half–rate receiver for a 10 Gbps automotive serial interface with 1–tap–unrolled 4–taps DFE and custom CDR algorithm. IEEE international symposium on circuits and systems (ISCAS) (May 27, 2018), pp. 1–5

References

12. A. Roshan Zamir, *High Speed Reconfigurable NRZ/PAM4 Transceiver Design Techniques: Doctoral Dissertation* (Texas A & M University, 2018), 100 p
13. H. Tang, L. Ding, J. Jin et al., A 28 Gb/s 2-tap FFE source-series-terminated transmitter in 22 nm CMOS FDSOI. 2018 IEEE international symposium on circuits and systems (ISCAS) (2018), pp. 1–4.
14. Universal Serial Bus 3.1 Specification, Revision 1.0 (2013), 631 p. https://manuais.iessanclemente.net/images/b/bc/USB_3_1_r1.0.pdf. Accessed 14 Mar 2019
15. PCI Express Base Specification Revision 3.0 (2010), 860 p. http://www.lttconn.com/res/lttconn/pdres/201402/20140218105502619.pdf. Accessed 14 Mar 2019
16. Serial ATA Revision 3.0 – Gold Revision (Serial ATA International Organization, 2009), 663 p. http://www.lttconn.com/res/lttconn/pdres/201005/20100521170123066.pdf. Accessed 14 Mar 2019
17. VESA DisplayPort Standard Version 1, Revision 1a (2008), 238 p. http://file.yizimg.com/383992/2014090921252964.pdf. Accessed 14 Mar 2019
18. https://www.synopsys.com/dw/ipdir.php?ds=dwc_ddr_multiphy
19. High-Definition Multimedia Interface Specification Version 1.4b (2011), 73 p. https://glenwing.github.io/docs/HDMI-1.4b.pdf. Accessed 14 Mar 2019
20. B. Razavi, *Design of Analog CMOS Integrated Circuits* (July, 2017), 801 p
21. P. Hakansson, A. Huynh, S. Gong, A study of wireless parallel data transmission of extremely high data rate up to 6.17 Gbps per channel. 2006 Asia-Pacific microwave conference (2006), pp. 975–978
22. E. Takahashi, N. Endou, Y. Kasai, M. Iwata, et al., 8 Gbps parallel data transmission with adaptive I/O circuit. IEEE Proc. 32nd Eur. Solid-State Circuits Conf. **32**(2), 472–475 (2006)
23. W. Xie, C. Guangqiang, J. Weiwei, Research on high-speed SerDes interface testing technology. 22nd international conference on electronic packaging technology (ICEPT) (2021), pp. 1–5
24. A. Sahakyan, A. Shishmanyan, A. Hekimyan, Multi-rate clock-data recovery solution in high speed serial links. Electronics and nanotechnology (ELNANO): IEEE 35th international conference (2015), pp. 242–244
25. C.L. Hsieh, S.L. Liu, A 1–16-Gb/s wide-range clock/data recovery circuit with a bidirectional frequency detector. IEEE Trans. Circuits Syst. II Express Briefs **58**, 487–491 (2011)
26. M.S. Li, Y.K. Lu, C.Y. Yang et al., PLL-based clock and data recovery for SSC embedded clock systems. 2019 international SoC design conference (ISOCC) (2015), pp. 309–310
27. M. El-Badry, M. El-Fiky, A. Yasser, A. Shehata et al., A 2.2-pJ/bit 10-Gb/s forwarded-clock serial-link transceiver for IoE applications. Signals, circuits and systems (ISSCS): International symposium (2017), pp. 1–4
28. J.W. Jung, B. Razavi, A 25-Gb/s 5-mWCMOS CDR/deserializer. VLSI circuits (VLSIC): Symposium (2012), pp. 138–139
29. M.M. Ayesh, S.A. Ibrahim, H.F. Ragai et al., A low-power high-speed charge-steering ADC-based equalizer for serial links. Electronics, circuits, and systems (ICECS): IEEE international conference (2015), pp. 500–501
30. O. Seijo, I. Val, J.A. Lopez-Fernandez et al., IEEE 1588 clock synchronization performance over time-varying wireless channels. 2018 IEEE international symposium on precision clock synchronization for measurement, control, and communication (ISPCS) (2018), pp. 1–6
31. S.H. Chung, Y.J. Kim, Y.H. Kim, et al., A 10-Gb/s 0.71-pJ/bit forwarded-clock receiver tolerant to high-frequency jitter in 65-nm CMOS. IEEE Trans. Circuits Syst. II Express Briefs **63**, 264–268 (2015)
32. X.. Wang, Q. Hu, Analysis and optimization of combined equalizer for high speed serial link. Anti-counterfeiting, security, and identification (ASID): IEEE 9th international conference (2015), pp. 43–46
33. M. Saneei, A. Afzali-Kusha, Z. Navabi, A mesochronous technique for communication in network on chips. 2006 international conference on microelectronics (2006), pp. 32–35

34. E. Kilada, M. Dessouky, A. Elhennawy, Architecture of a fully digital CDR for plesiochronous clocking systems. 2007 IEEE international conference on signal processing and communications (2007), pp. 939–942
35. S.S. Saber, M. Ehsanian, A linear high capture range CDR with adaptive loop bandwidth for SONET application. Microelectronics (ICM): 29th international conference (2017), pp. 1–4
36. J. Kim, Y. Hwang, Y. Moon, A study of the referenceless CDR based on PLL. SoC design conference (ISOCC): IEEE international conference (2016), pp. 265–266
37. H. Zhang, P. Xue, Z. Hong, A 4.6–5.6 GHz constant KVCO low phase noise LC-VCO and an optimized automatic frequency calibrator applied in PLL frequency synthesizer. 43rd annual conference of the IEEE Industrial Electronics Society (2017), pp. 8337–8342
38. J. Jin, J. Kim, H. Kim et al., A 4.0-10.0-Gb/s referenceless CDR with wide-range, jitter-tolerant, and harmonic-lock-free frequency acquisition technique. ESSCIRC 2018-IEEE 44th European solid state circuits conference (ESSCIRC) (2018), pp. 146–149
39. P. Zhang, C. Zhang, J. Zhang et al., A 25–28Gb/s PLL-based full-rate reference-less CDR in 0.13 μm SiGe BiCMOS. 2017 2nd IEEE international conference on integrated circuits and microsystems (ICICM) (2017), pp. 186–190
40. Y. He, Z. Wang, H. Liu et al., An 8.5–12.5 ghz multi-pll clock architecture with lc PLL and ring PLL for multi-lane multi-protocol SerDes. 2017 international conference on electron devices and solid-state circuits (EDSSC) (2017), pp. 1–2
41. N. Bansal, R. Gupta, An NMOS low drop-out voltage regulator with-17dB wide-band power supply rejection for SerDes in 22FDX. 31st international conference on VLSI design and 2018 17th international conference on embedded systems (VLSID) (2018), pp. 341–346
42. E. Abramov, T. Vekslender, O. Kirshenboim, et al., Fully integrated digital average current-mode control voltage regulator module IC. IEEE J. Emerg. Select. Topics Power Electron., 485–499 (2017)
43. S. Harutyunyan, H. Kostanyan, M. Grigoryan, et al., A reliable PMOS-based charge pump architecture. Proc. RA NPUA Ser. Tech. Sci **73**(2), 181–187 (2020)
44. J. Cui, Y. Zeng, J. Xia, Design of a low temperature drift and high PSRR bandgap reference source with second-order compensation. J. Terahertz Sci. Electron. Inf. Technol., 41–49 (2018)
45. X. Xu, C. Chen, T. Sugiura et al., 18-GHz band low-power LC VCO IC using LC bias circuit in 56-nm SOI CMOS. IEEE Asia Pacific microwave conference (APMC) (2017), pp. 938–941
46. X. Xu, C. Chen, T. Yoshimasu et al., A 28-GHz band highly linear power amplifier with novel adaptive bias circuit for cascode MOSFET in 56-nm SOI CMOS. International conference on electron devices and solid-state circuits (EDSSC) (2017), pp. 1–2
47. I.M. Filanovsky, M. Igor, A. Allam et al., Sub-regulators for biasing circuits. IEEE 55th international midwest symposium on circuits and systems (MWSCAS) (2012), pp. 314–317
48. X. Xu, X. Yang, T. Yoshimasu, A 2-GHz-band low-phase-noise VCO IC with an LC bias circuit in 180-nm CMOS. 11th European microwave integrated circuits conference (EuMIC) (2016), pp. 197–200
49. D. Chen, A 16b 5MSPS two-stage pipeline ADC with self-calibrated technology. International conference on information and computer technologies (ICICT) (2018), pp. 155–158
50. H. Alaqil, J. Hong, Combined microstrip and suspended substrate stripline combline bandpass filter with two transmission zeros. 15th Mediterranean microwave symposium (MMS) (2015), pp. 1–4
51. M. Sarkar, Suspended substrate stripline-microstrip mixed substrate topology based wide stopband low pass filter. TEQIP III sponsored international conference on microwave integrated circuits, photonics and wireless networks (IMICPW) (2019), pp. 90–94
52. G. Xiang, K. Sheach, P. Brunet, A study on high-density high-speed SerDes design in buildup flip chip ball grid array packages. European microelectronics and packaging conference (2019), pp. 1–4

53. N. Na, J. Audet, L. Shan, Design optimization for isolation in high wiring density packages with high speed SerDes links. 56th electronic components and technology conference (2006), pp. 1–7
54. T. Tang, B.S. Fang, D. Ho et al., Innovative flip chip package solutions for automotive applications. IEEE 69th electronic components and technology conference (ECTC) (2019), pp. 1432–1436
55. V. King, S. Jared, *Transmission-Line Theory* (1955), 509 p
56. C.R. Paul, *Analysis of Multiconductor Transmission Lines* (2007), 414 p
57. F.H. Branin, Transient analysis of lossless transmission lines. Proc. IEEE **55**, 2012–2013 (1967)
58. T. Itoh, C. Caloz, *Electromagnetic Metamaterials: Transmission Line Theory and Microwave Applications* (Wiley, 2005), 352 p
59. V.G. Oklobdzija R.K. Krishnamurthy, *High-Performance Energy-Efficient Microprocessor Design* (Springer Science & Business Media, 2006), 338 p
60. O.H. Petrosyan, A.A. Martirosyan, A.S. Trdatyan, et al., Equalization method of resistors. Manual Eng. Acad. Armenia Yerevan **15**(3), 475–479 (2018)
61. V.Sh. Melikyan, A.K. Hayrapetyan, B.E. Baghramyan et al., Transmitter output impedance calibration method. Proceedings of IEEE East-West design & test symposium (EWDTS) (Kazan, Russia, 2018), pp. 51–58
62. V. Melikyan, A. Balabanyan, A. Hayrapetyan et al., Receiver/transmitter input/output termination resistance calibration method. 2013 IEEE XXXIII international scientific conference electronics and nanotechnology (ELNANO) (2013), pp. 126–130
63. Z. Yan, C. Zhang, M. Wang et al., Calibration mechanism for input/output termination resistance in 28 nm CMOS. IEEE 3rd international conference on integrated circuits and microsystems (ICICM) (2018), pp. 42–46
64. R.G. Chambers, The anomalous skin effect. Proc. R. Soc. Lond. A. Math. Phys. Sci. **215**, 481–497 (1952)
65. E.J. Murphy, H.H. Lowry, The complex nature of dielectric absorption and dielectric loss. J. Phys. Chem. **34**, 598–620 (2002)
66. M. Tang, Z. Li, J. Hu et al., A 56-Gb/s PAM4 continuous-time linear equalizer with fixed peaking frequency in 40-nm CMOS. IEEE international conference on integrated circuits, technologies and applications (ICTA) (2019), pp. 89–90
67. M.S. Choudhary, N.S. Pudi, J.M. Redoute et al., An EMI immune PAM4 transmitter in 130 nm BiCMOS technology. IEEE MTT-S international microwave and RF conference (IMARC) (2019), pp. 1–4
68. C. Menolfi, T. Toifl, R. Reutemann et al., A 25 Gb/s PAM4 transmitter in 90 nm CMOS SOI. 2005 IEEE international digest of technical papers. Solid-state circuits conference (ISSCC) (2005), pp. 72–73
69. J. Li, S. An, Q. Zhu, et al., VSB modified duobinary PAM4 signal transmission in an IM/DD system with mitigated image interference. IEEE Photon. Technol. Lett. **32**(7), 363–366 (2020)
70. Q. Liao, N. Qi, Z. Zhang et al., The design techniques for high-speed PAM4 clock and data recovery. IEEE international conference on integrated circuits, technologies and applications (ICTA) (2018), pp. 142–143
71. R. Ma, M. Cao, G. Chen et al., A 5/10 Gb/s dual-mode NRZ/PAM4 CDR in 65-nm CMOS. IEEE international conference on electron devices and solid-state circuits (EDSSC) (2019), pp. 1–3
72. M. Wang, Y. Chen, J. Yuan, A low jitter 50 Gb/s PAM4 CDR of receiver in 40 nm CMOS technology. International conference on wireless communications and signal processing (WCSP) (2020), pp. 349–352
73. D.M. Pozar, *Microwave Engineering* (Wiley, 2011), 736 p
74. R.N. Bracewell, R.N. Bracewell, *The Fourier Transform and Its Applications* (1986), 640 p
75. V. Serov, *Fourier Series, Fourier Transform and Their Applications to Mathematical Physics* (2017), 534 p

76. J. He, N. Qi, N. Yu et al., A 2nd-order CTLE in 130 nm SiGe BiCMOS for a 50 GBaud PAM4 Optical Driver. IEEE international conference on integrated circuits, technologies and applications (ICTA) (2018), pp. 151–152
77. U. Upadhyaya, S. Sen, S. Goyal et al., A 16 Gbps 10: 1 serializer with active inductor based CTLE for high frequency boosting. 27th IEEE international conference on electronics, circuits and systems (ICECS) (2020), pp. 1–4
78. A. Mkhitaryan, A. Grigoryan, M. Grigoryan, et al., Hysteresis improvement method in MIPI D-PHY low-power receiver. Manual of Russian-Armenian Slavonic University. Phys. Math. Natural Sci. **1**, 95–103 (2020)
79. P. Hale, J. Jargon, C. Wang, et al., A statistical study of de-embedding applied to eye diagram analysis. IEEE Trans. Instrum. Measur. **61**(2), 475–488 (2011)
80. M. Mehri, R. Sarvari, A. Seydolhosseini, Eye diagram parameter extraction of nano scale VLSI interconnects. IEEE 21st international conference on electrical performance of electronic packaging and systems (EPEPS) (2012), pp. 327–330
81. B. Gao, K. Wei, L. Tong, An eye diagram parameters measurement method based on K-means clustering algorithm. IEEE MTT-S international microwave symposium (IMS) (2019), pp. 901–904
82. P. Li, T. Wu, An eye diagram improvement method using simulation annealing algorithm. IEEE 22nd workshop on signal and power integrity (SPI) (2018), pp. 1–4
83. J. Park, D. Kim, Y. Kim et al., Eye-diagram estimation with stochastic model for 8B/10B encoded high-speed channel. IEEE international symposium on electromagnetic compatibility and 2018 IEEE Asia-Pacific symposium on electromagnetic compatibility (EMC/APEMC) (2018), pp. 1–5
84. D. Tonietto, J. Hogeboon, E. Bensoudane et al., A 7.5 Gb/s transmitter with self-adaptive FIR. 2008 IEEE symposium on VLSI circuits (2008), pp. 198–199
85. P. Prandoni, M. Vetterli, *Signal Processing for Communications* (2008), 371 p
86. R. Marin, A. Frappé, A. Kaiser, Delta-sigma based digital transmitters with low-complexity embedded-FIR digital to RF mixing. IEEE international conference on electronics, circuits and systems (ICECS) (2016), pp. 237–240
87. H. Liu, H. Jiang, Y. Shen et al., A delta-sigma-based transmitter utilizing FIR-embedded digital power amplifiers. IEEE 58th international midwest symposium on circuits and systems (MWSCAS) (2015), pp. 1–4
88. E. Conde-Almada, Design and physical implementation of an analog receiver for a SerDes system on chip in 130nm CMOS technology (2016), 48p
89. T. Terada, R. Fujiwara, G. Ono et al., A CMOS UWB-IR receiver analog front end with intermittent operation. IEEE symposium on VLSI circuits (2007), pp 86–87
90. A. Jaiswal, Y. Fang, K. Hofmann, Low-power high-speed on-chip asynchronous wave-pipelined CML SerDes. 27th IEEE international system-on-chip conference (SOCC) (2014), pp. 5–10
91. A. Aghighi, A. Tajalli, M. Taherzadeh-Sani, A low-power 10 to 15 Gb/s common-gate CTLE based on optimized active inductors. IFIP/IEEE 28th international conference on very large scale integration (VLSI-SOC) (2020), pp. 171–175
92. B. Li, B. Jiao, C. Chou, et al., CTLE adaptation using deep learning in high-speed SerDes link. IEEE 70th electronic components and technology conference (ECTC) (2020), pp. 952–955
93. V. Melikyan, A. Sahakyan, et al., High PSRR and accuracy receiver active equalizer. IEEE 34th international scientific conference on electronics and nanotechnology (ELNANO) (2014), pp. 194–197
94. P. Francese, T. Toifl, M. Brändli et al., A 16 Gb/s receiver with DC wander compensated rail-to-rail AC coupling and passive linear-equalizer in 22 nm CMOS. ESSCIRC 2014-40th European solid state circuits conference (ESSCIRC) (2014), pp. 435–438
95. M. Chen, M. Chung, C. Yang, A low-PDP and low-area repeater using passive CTLE for on-chip interconnects. Symposium on VLSI circuits (VLSI circuits) (2015), pp. 244–245

References

96. D. Thulasiraman, G. Chiranjeevi, J. Gaggatur et al., A 18.6 fJ/bit/dB power efficient active inductor-based CTLE for 20 Gb/s high speed serial link. IEEE international conference on electronics, computing and communication technologies (CONECCT) (2019), pp. 1–6
97. Y. Choi, Y. Kim, A 10-Gb/s receiver with a continuous-time linear equalizer and 1-tap decision-feedback equalizer. IEEE 58th international midwest symposium on circuits and systems (MWSCAS) (2015), pp. 1–4
98. J. Chae, M. Kim, S. Choi, et al., A 10.4-Gb/s 1-tap decision feedback equalizer with different pull-up and pull-down tap weights for asymmetric memory interfaces. IEEE Trans. Circuits Syst. II Express Briefs **67**(2), 220–224 (2019)
99. K. Kaviani, A. Amirkhany, C. Huang, et al., A 0.4-mW/Gb/s near-ground receiver front-end with replica transconductance termination calibration for a 16-Gb/s source-series terminated transceiver. IEEE J. Solid-State Circuits **48**, 636–648 (2013)
100. W. Fu, Q. Hu, R. Wang, A 40 Gb/s PAM4 SerDes receiver in 65 nm CMOS technology. IEEE Canadian conference on electrical & computer engineering (CCECE) (2018), pp. 1–4
101. L. Tang, W. Gai, L. Shi, PAM4 receiver with adaptive threshold voltage and adaptive decision feedback equalizer. 2016 IEEE international symposium on circuits and systems (ISCAS) (2016), pp. 2246–2249
102. B. Mrković, M. Ašenbrener, The simple CMOS negative capacitance with improved frequency response. 2012 proceedings of the 35th international convention MIPRO (2012), pp. 87–90
103. M. Grigoryan, A. Atanesyan, G. Hakobyan, S. Harutyunyan, Two stage CTLE for high speed data receiving. IEEE 40th international conference on electronics and nanotechnology (ELNANO) (Kyiv, 2020), pp. 374–377
104. E.V. Balashov, D. Pasquet, A.S. Korotkov et al., Automatization of compression point 1dB (CP1dB) and input 3rd order intercept point (IIP3) measurements using lab VIEW platform. International symposium on signals, circuits and systems (2005), pp. 195–198
105. W. Kester, A.D.I. Engineeri, *Data Conversion Handbook* (2005), 976 p
106. A.A. Atanesyan, M.T. Grigoryan, H.V. Margaryan, H.A. Aghayan, et al., Method of increasing current DAC linearity with considering its random variables for modeling risk or uncertainty. Manual of Russian-Armenian (Slavic) University. Phys. Math. Natural Sci. **2**, 64–70 (2020)
107. H. Aghayan, D. Manukyan, M. Grigoryan, Low input capacitance dynamic latch comparator for high speed operation. Manual of Russian-Armenian (Slavic) University. Phys. Math. Natural Sci. **1**, 65–75 (2020)
108. https://visualstudio.microsoft.com/
109. V.Sh. Melikyan, A.A. Atanesyan, M.T. Grigoryan, H.T. Kostanyan et al., Duty-cycle correction circuit for high speed interfaces. IEEE 39th international conference on electronics and nanotechnology (ELNANO) (Kyiv, 2019), pp. 42–45
110. V. Melikyan, A. Trdatyan, A. Sahakyan, A. Martirosyan et al., Process variation detection and self-calibration method for high-speed serial links. IEEE East-West design & test symposium (EWDTS), 14 September 2018 (Kazan, Russia, 2018), pp. 681–684

Chapter 2
Design Methods of Integrated Circuits, Working Under Non-standard Operating Conditions

2.1 General Issues of Design Methods of Integrated Circuits, Working Under Non-standard Operating Conditions

2.1.1 Need for Design Methods of Integrated Circuits, Working Under Non-standard Operating Conditions

It is known that over the past 60 years, the semiconductor industry has evolved according to Moore's Law [1], which has allowed to have metal-oxide-semiconductor (MOS) transistors with channel lengths up to 3 nm [2]. All this has led to an increase in the productivity of ICs, because the small size of transistors allows to have more functional blocks on the same area. Along with all this, the operating frequencies of ICs have also increased, reaching tens of GHz [3–5]. Such data transfer speeds have allowed the development and improvement of such technological directions and industries as automotive electronics [6], Internet of Things [7], artificial intelligence [8], and virtual reality devices [8]. This ensures continuous data transfer without human-human or human-computer interaction [9]. During the simulation stage, there is a need to observe the phenomena of voltage and temperature drift [10], because during the operation of ICs, drastic changes in external conditions are possible even after the calibration stage [11]. Such improvements, as well as the greater involvement of high-speed devices in human life, have led to the tightening of requirements for IC reliability and the need to develop means of designing ICs that work in non-standard operating conditions. Disruption of IC operation can directly or indirectly lead to serious economic losses, environmental damage, or threats to human life.

The reduction in sizes led to thinning of the oxide layer of transistor gates, which allowed to reduce their threshold voltage [11, 12]. The low threshold voltage, in turn, allowed to reduce the supply voltages of ICs, leading to the reduction of power consumption [13–16]. The development of the technological process has made it

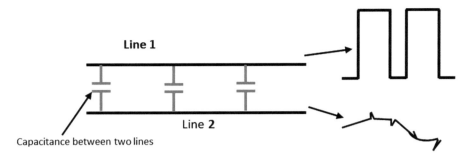

Fig. 2.1 Capacitance between the lines

possible to reduce the distance between metal layers, leading to an increase in the density of interconnects [17, 18]. In the case of closer wires, the potential connections between them increase [19]. Large capacitances lead to an increase in the noise level (Fig. 2.1) [20–23]. In modern circuits, the value of the supply voltage has become several hundred millivolts, and noises reach tens of millivolts [24–26].

An increase in noise and a decrease in the value of the supply voltage led to a decrease in the signal-to-noise ratio. This reduces the noise immunity of modern ICs [27, 28] and can lead to undesirable changes in the voltage characteristics of circuits, up to functional failure.

The increase in operating frequencies, the density of transistors and interconnects per unit area, and the reduction of the thickness of the oxide layer have made the influence of the undesirable phenomena of self-heating [29] and aging [30] on the operation of ICs more tangible. Under the influence of self-heating and electric field, the charge carriers can acquire such kinetic energy that will allow them to appear in the oxide layer of the gate, being absorbed in it. This process is known as the phenomenon of hot carriers injection (HCI) [31], the effect of which is great during switching of transistors (Fig. 2.2) [32]. The phenomenon of bias temperature instability (BTI) occurs when a constant voltage is applied to the terminals of the transistor (Fig. 2.2) [32].

A high potential difference between the gate and the rest of the terminals causes damage to the oxide layer, which is recoverable over time. The continuity of BTI and HCI phenomena leads to a change in the properties of the oxide layer of transistors and its parameters, as well as a decrease in the lifetime of ICs [33–35]. Modern standards require ICs to meet a minimum 10-year life span limitation, which further increases the role of aging phenomena. Factory-provided transistor aging models are used to assess the effect of BTI and HCI phenomena on circuit performance. As a result of the simulation, changes in the threshold voltage and current value of transistors in the saturation region are observed [36, 37]. Taking into account the fields of application of current ICs, as well as the operating temperature of automotive electronics standards up to 150 °C and a larger range of offset value [38–41], the mentioned phenomena can play a decisive role from the point of view of IC functionality.

2.1 General Issues of Design Methods of Integrated Circuits, Working... 61

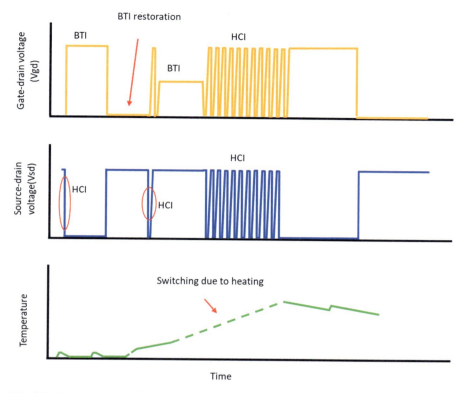

Fig. 2.2 Occurrence of HCI, BTI, self-heating phenomena during the operation of the transistor

Along with the reduction of the supply voltages, the amplitudes of the signals applied to the inputs of the comparators also decreased [42]. Systems with a small amplitude compared to the constant component of the voltage require operational amplifiers (OpAmp) with high sensitivity and gain. The sensitivity is characterized by the minimum difference between inputs required for switching. Ideally, for any voltage difference between the inputs greater than 0, the output of the comparator should be switched. The minimum voltage difference between the inputs, at which switching occurs at the output of the comparator, is called the offset voltage (Fig. 2.3) [43, 44].

The main causes of offset are errors in technological process, asymmetry of layout, and aging phenomena [45–47].

In modern ICs, the value of offset reaches tens of millivolts, which is a challenge for design companies. The circuit of the amplifier with a folded cascode belongs to the series of well-known structures providing a high value of gain in comparators [48] (Fig. 2.4).

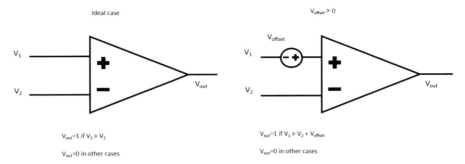

Fig. 2.3 The input bias voltage in the comparators

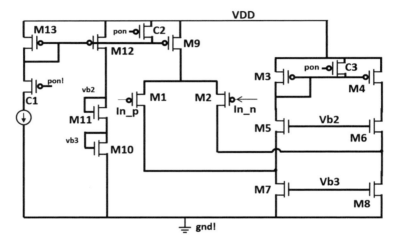

Fig. 2.4 View of the OpAmp with folded cascode

The simulation results show the presence of offset voltage up to 35 mV due to aging of the OpAmp in the off state in the case of different scenarios of input signals (Table 2.1).

It follows from the above that the increase in operating frequencies and noises and the development of the technological process have led to a number of complications in IC design process. Taking into account the greater involvement of ICs in human life every year, the possibility of drastic changes in operating conditions, and the strict technical requirements for ICs, the development of ICs operating in non-standard conditions becomes a challenge. There is a need to develop new methods of designing ICs that operate under non-standard conditions, which will reduce the effect of offset of input signals in comparators, self-heating, and aging phenomena on the operation of circuits, increasing the noise immunity and efficiency of digital and analog cells.

2.1 General Issues of Design Methods of Integrated Circuits, Working...

Table 2.1 OpAmp input offset due to off-state aging

Offset voltage (mV)	Scenario M1/M2
25	0/1
35	0/0
18	1/1
25	1/0

2.1.2 Existing Design Methods of Integrated Circuits, Working Under Non-standard Operating Conditions

In order to solve the problems mentioned in the previous sub-chapter, new methods and means have been developed in recent years, the use of which contributes to increasing the stability and uninterrupted operation of ICs. It was possible to achieve all this by introducing digital subsystems that ensure self-regulation of circuits. The use of digital-to-analog and analog-to-digital converters allows to introduce calibration algorithms [49], which, taking into account the operating conditions of the blocks, can ensure the optimal values of resistances, current, voltage, and other parameters at individual nodes of circuits by changing the code. Similar algorithms make it possible to achieve the minimum change of gain, delay, signal jitter, and other important parameters of the circuit. Among the mentioned solutions are the methods of reducing the offset with digital control [50] and automatic reset in comparators [51] in IC receivers, as well as the method of continuous data reception through digital delay lines [52–55].

The Method of Reducing Offset in Comparators

It is known that offset is one of the important factors limiting the accuracy of comparators and OpAmps [56]. Used primarily in analog circuits, these elements are one of their most sensitive components. One of the complications in the design of comparators is the provision of saturation and high gain of circuit transistors and symmetry of layout [57]. These three factors have quite a large influence on the value of the offset. An unbalanced layout can cause differences in resistance and capacitance values between comparator branches, leading to an increase in offset [58–60].

The two-cascade comparators used in current ICs are able to provide a low offset value, having a low supply voltage value and high noise immunity [56]. A small value of offset voltage can be achieved for two reasons:

1. The offset of the first cascade (Fig. 2.5) [56] is canceled by the input offset storage circuit.
2. The offset of the second cascade (Fig. 2.6) [56] is reduced due to the high gain of the first cascade.

The first cascade is a dynamic amplifier that amplifies the difference of the input signals by discharging the differential junction [56].

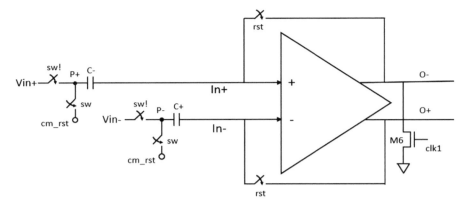

Fig. 2.5 The first cascade of the comparator

Fig. 2.6 The second cascade of the comparator

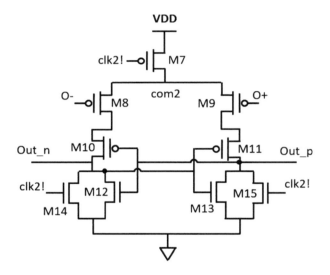

The second cascade represents another dynamic amplifier connected to the switch [56].

The outputs of the first cascade feed the second cascade, which, due to the positive feedback of the flip-flop, generates a logic zero or one at the output. The input offset storage circuit consists of sw and rst control signals and C+/C- capacitances. M7 transistor functions both as a switch and as a current source to increase the gain of the second cascade. The principle of operation of the comparator (Fig. 2.7) [56] is divided into two parts: reset and comparison. The offset voltage cancellation is done in the first stage.

Four control signals (rst, clk1, clk2, sw) are used to ensure the operation of the comparator at different stages. The clk2 signal is responsible for the comparison stage, and the other three are responsible for canceling the offset. First, the input

2.1 General Issues of Design Methods of Integrated Circuits, Working... 65

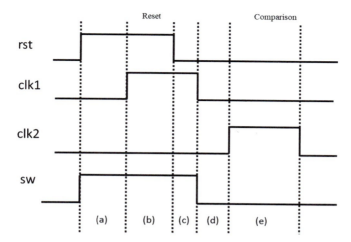

Fig. 2.7 Timing diagram of comparator operation. (**a**) Charging of C+/C- capacitors up to VDD value. (**b**) Discharge of capacitors. (**c**) Offset storage. (**d**) Outputs of the first cascade are connected to the VDD point and input signals are sampled. (**e**) Comparison of signals

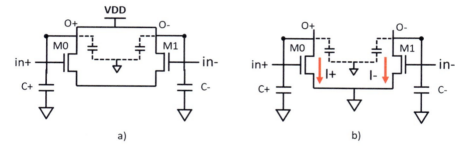

Fig. 2.8 A simplified view of the operation of the comparator in the reset phase

switches are connected through the rst signal. The In+ and In- points are connected to the O+ and O- points, respectively, and are charged up to the VDD value of the supply voltage (Fig. 2.8a) [56]. The clk1 signal turns on the M6 transistor, discharging C+/C- potentials (Fig. 2.8b) [56]. As the input M0 and M1 transistors pass into the subthreshold range, the I+/I- currents decrease sharply. The rst signal is turned off, and the offset is stored in C+/C- capacitors (Fig. 2.9a) [56].

O+ and O- points are connected to VDD; at the same time the sw signal is turned off to register input signals (Fig. 2.9b) [56]. A comparison of input signals is made through the discharge of O+ and O- points (Fig. 2.9c) [56].

Then, the O+ and O- points, connected to the inputs of the second cascade, form the outputs of the general system (Fig. 2.10) [56]. The main reason for the offset of the first cascade is the asymmetry of M1 and M2 transistors. It is canceled during the reset phase. During the discharge process, the gate and output of input transistors are connected together. Therefore, under the influence of the current flowing through

66 2 Design Methods of Integrated Circuits, Working Under Non-standard...

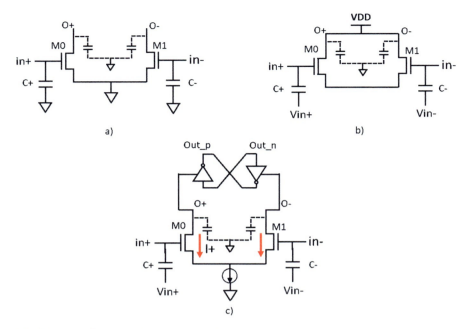

Fig. 2.9 Simplified view of the operation of the comparator during the comparison stage

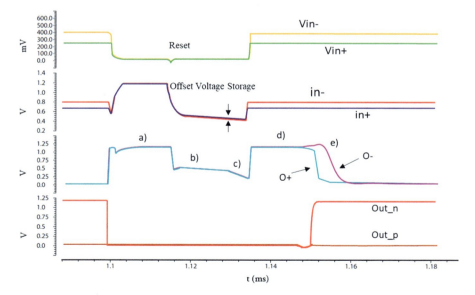

Fig. 2.10 Process of operation of a comparator

2.1 General Issues of Design Methods of Integrated Circuits, Working...

them, the voltage applied to the gate will increase, which in turn will cause it to shrink. Due to such negative feedback, the current values of the input transistors will get closer to each other.

After the input transistors pass into the subthreshold range, the difference in these currents is less than 1 nA. After the rst signal is turned off, the value of the current for the input transistors has the following dependence on the gate-source voltage (2.1) [56]:

$$I_{ds} = 2n\beta U_T^2 e^{\frac{V_G - V_{th}}{nU_T}} \qquad (2.1)$$

where I_{ds} is the value of the current flowing through the input transistors, n is the slope coefficient of the current graph in the subthreshold range, V_{gs} is the gate-source voltage, V_{th} is the threshold voltage of the transistor, and U_T (2.2) and β (2.3) are the current and thermal voltage coefficients, respectively;

$$\beta = \mu C_{ox} \frac{W}{L} \qquad (2.2)$$

$$U_T = \frac{kT}{q}, \qquad (2.3)$$

where μ_n is the charge carrier mobility, C_{ox} is the capacitance per unit area of the transistor gate insulator, W and L are the width and length of the transistor channel, k is the Boltzmann constant, and T is the absolute temperature [56, 61]. After the reset phase, the voltage stored in C+/C- potentials is determined by the expression given in (2.4) [57]:

$$V_g \sim V_{th} + nU_T \ln\left(\frac{I_{ds}}{2n\beta U_T^2}\right). \qquad (2.4)$$

From Eq. (2.4), it follows that asymmetries arising from β and Vth are also stored and canceled before the comparison stage.

The Method of Offset Reduction Using a Digital Control Circuit in Receivers of Integrated Circuits

Advances in IC manufacturing, as well as innovative engineering solutions, have made it possible to have data transmission systems reaching tens of GHz [62, 63]. Such systems require efficient communication between multiple ICs for accurate operation of the entire system. For this, high-speed transmitter-channel-receiver (TCR) (Fig. 2.11) [63] systems of data transfer were developed. At the transmitter node, the parallel-to-serial data conversion (PSC) circuit converts the incoming parallel data to serial data for transmission over the channel. At the output of the receiver, the data is parallelized by means of a serial-to-parallel conversion

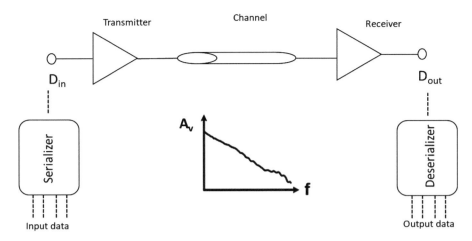

Fig. 2.11 TCR system

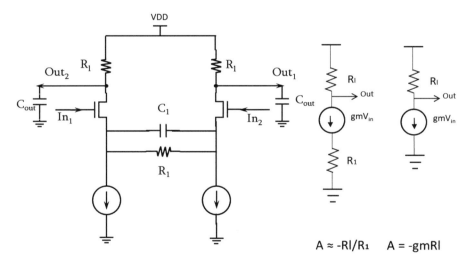

Fig. 2.12 Input data equalizer of a receiver and a simplified view of circuit operation

(SPC) circuit. In the ideal case, the channel represents a short circuit between transmitters and receivers [63].

The non-ideality of channel parameters—low bandwidth [64] and noises [65]—affect the accuracy of data reception at the input of the receiver. As it is well known, the transmission characteristic of the channel has a low-frequency character [66], which leads to distortions of the transmitted data depending on their frequency. To solve the mentioned problems, data equalizers [67–72] are used in both transmitters and receivers, which have opposite characteristics to the channel in order to neutralize its losses. The equalizer used at the input of the receiver (Fig. 2.12) [67]

2.1 General Issues of Design Methods of Integrated Circuits, Working...

consists of a differential pair, the two branches of which are connected to each other by resistors and capacitors. The feature of the circuit is that depending on the frequency of the input data, the gain takes different values. At low frequencies, the behavior of the circuit is similar to that of a source-degraded amplifier, and at high frequencies, a common-source amplifier.

Having the values of R_b, R_1 resistors and C_e, C_1 capacitors, the system gain (2.5) can be obtained.

$$A_V = \frac{g_m}{Cr} \frac{S + \frac{1}{R_o C_o}}{\left(S + \frac{1+\frac{g_m R_1}{2}}{R_1 C_1}\right)\left(S + \frac{1}{R_o C_o}\right)}, \qquad (2.5)$$

where s is a complex variable and gm is the conductance of the transistor.

From the amplitude-frequency characteristic of the system obtained as a result of the simulation, it is clear that the circuit weakens the low and strengthens the high frequencies (Fig. 2.13). Thus, the system, having a characteristic opposite to the channel, is able to provide equalized data at the input of the receiver. In order to reduce the level of signals to logic zero and one, one or more cascade amplifiers are used after the equalizer [73, 74].

To estimate the degree of signal distortion, the eye diagram is used. It is obtained by cutting the data signal according to the clock signal and superimposing these parts on each other (Fig. 2.14) [75].

The distorted data signal causes the eye diagram to close. A small gap in the diagram leads to a smaller storage of signal timing parameters: jitter, duration, and amplitude of rise/falls.

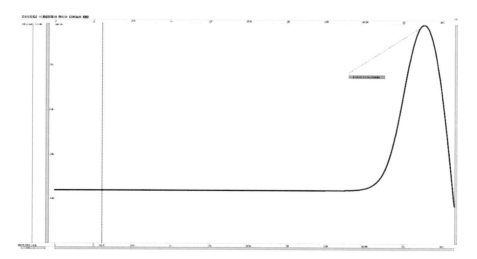

Fig. 2.13 Amplitude-frequency characteristic of an equalizer

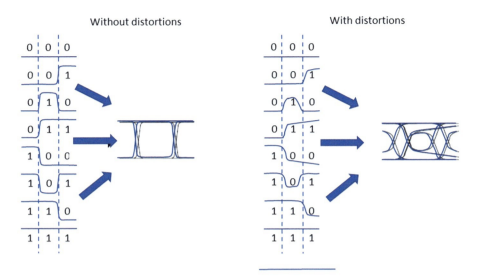

Fig. 2.14 Obtaining an eye diagram

The presence of an offset between the output signals of a node can lead to incorrect switching or loss of bits. The value of the offset voltage can be determined by calculating the difference between the outputs of the circuit when the input signals are equal. In modern ICs, analog-digital self-regulating blocks are used to reduce the offset between the outputs of the equalizer [76, 77]. Their purpose is to provide a minimum voltage difference between the outputs by changing the code for an input single-ended signal. The most well-known method consists of a DAC and a differential pair, which, by connecting to the outputs of the equalizer, reduces the offset due to the insertion of additional current (Fig. 2.15).

DAC represents resistors connected in series and MOS switches controllable by code (Fig. 2.16). The latter, depending on the digital code and the reference voltage value [78, 79], output the analog voltage caused by the connection of resistors. By increasing the bit-rate of the DAC, it is possible to increase the accuracy of the system at the expense of reducing the step of the analog voltage [79]. The resulting analog voltage is supplied to M1 and M2 transistors (Fig. 2.15), whose connected current source can either match the temperature-independent current supplied to DAC or differ from it by being connected to a circuit with a constant conductivity to keep the gain stable during the operation.

The calibration algorithm works in the following sequence: at the beginning point b) (Fig. 2.18), being connected to the point providing the smallest voltage of DAC, is kept constant. The code changes from maximum to minimum and vice versa (Fig. 2.18). During that time, the voltage of point a) starts to decrease from the high voltage value of DAC in a minimal step, equaling to point b) (Fig. 2.17).

Fig. 2.15 Offset reduction circuit of an equalizer

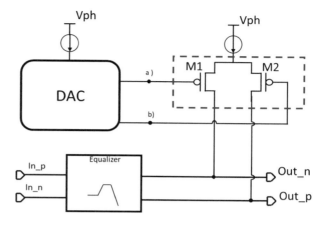

Fig. 2.16 DAC circuit

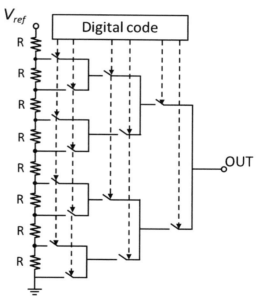

In the next stage, with the same logic, the voltage of point b) increases and decreases from the beginning, keeping the voltage of point a) constant. At the end, the voltage of point a) increases, keeping point b) fixed (Fig. 2.18).

During calibration, the inputs of the node are connected to each other so that it is possible to measure the offset. The required code value is selected during output switching. After that, the calibration is considered complete, and sequential data can be accepted.

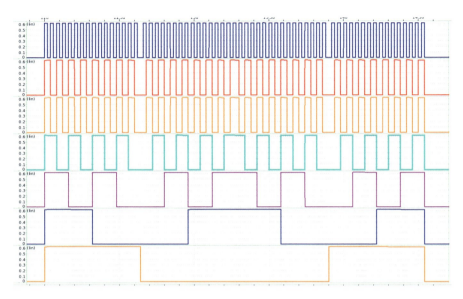

Fig. 2.17 The sequence of changing DAC code during calibration

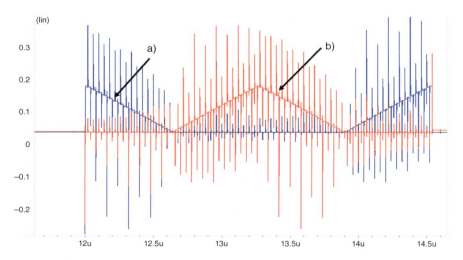

Fig. 2.18 The process of changing the values of points (**a**) and (**b**) during calibration

A Method of Continuously Receiving Sequential Data Using Digital Delay Lines

One of the most important parameters of modern ICs is performance [80]. Taking the current operating frequencies into account, the requirement to have a clock signal with a stable frequency inside ICs increases [81]. Increasing operating frequencies

2.1 General Issues of Design Methods of Integrated Circuits, Working... 73

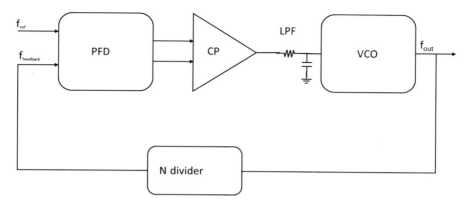

Fig. 2.19 A simplified block diagram of a PLL

has led to shorter signal periods, which means that the negative impact of signal jitter and drift during data transmission has increased. In modern systems, as a generator of a constant clock signal, the PLL is used (Fig. 2.19). It is a self-calibrating system, the output frequency of which is a multiple of the input frequency. Powered by a clock generator, the frequency of which is several tens of megahertz, PLL allows to have a clock signal reaching tens of GHz in ICs. The phase-frequency detector (PFD) included in its structure is intended for matching the phases of input and output signals. By registering the phase difference, it generates output voltages that control the charge pump (CP). Depending on the output voltage of the CP, the voltage-controlled oscillator (VCO) changes the frequency of the output signal. The ratio of the frequencies of the input and output signals is determined by means of a frequency divider (FD) (Fig. 2.19).

Such a structure of the clock signal generator occupies a large area and does not allow to use it in different parts of ICs [82]. The clock signal is transmitted to other parts of the IC by buffering (Fig. 2.20) [83].

The main purpose of buffering is to restore the signal from distortion. Connections, having their own parasitic resistances and making capacitive connections with neighboring wires, slow down the signal rise/falls, affect the duty cycle, and can cause jitter and signal delay [84] (Fig. 2.21). This interferes with the parallel operation of different parts of ICs.

Elements used for signal buffering contain simplest inverters in their structure [85]. Depending on the conditions of the external environment, the timing characteristics of inverters can be changed, because they affect the operation of transistors. Temperature fluctuations lead to a change in the threshold voltage of transistors, which affects the value of the charge and discharge current of the output load of inverters. Fluctuations in the supply voltage, affecting the value of the gate-source voltage, also lead to a change in timing parameters of inverters.

In order to solve the mentioned problems in modern ICs, programmable digital delay lines (DDL) are used (Fig. 2.22) [86, 87]. Consisting of sequentially connected

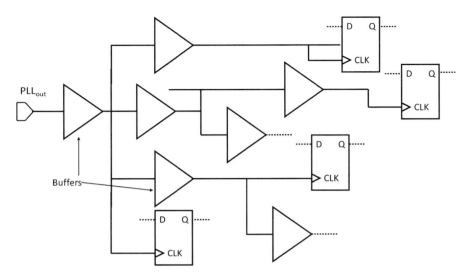

Fig. 2.20 Buffering of PLL output clock signal

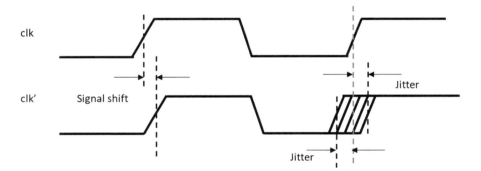

Fig. 2.21 A simplified block diagram of a PLL

delay buffers, the circuit allows code to control the overall line delay by enabling or disabling the appropriate number of elements.

Similar circuits are also used for uninterrupted reception of sequential data. It is possible to control the clock signal using DDLs. Depending on the external conditions, the calibration algorithm selects the appropriate code, placing the rise/fall of the clock signal in the middle of the data signal, for the most accurate sampling. Thanks to shifting of the clock signal, it is possible to sample the data with both rise and fall (Fig. 2.23). The most important parameters of DDLs are the total delay interval of the line and the system step per code unit change.

Table 2.2 shows the simulation results for the main DDL parameters.

2.1 General Issues of Design Methods of Integrated Circuits, Working...

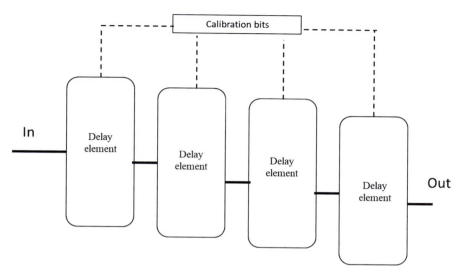

Fig. 2.22 DDL block diagram

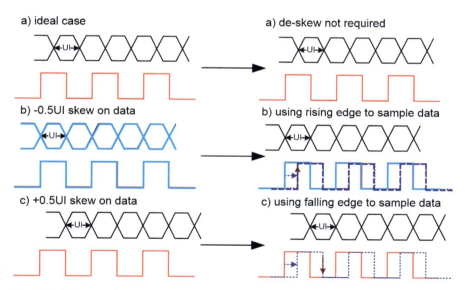

Fig. 2.23 DDL operation principle

The maximum step bias from the standard value is 0.03 MM, which is 1.5% of the signal period. The total delay range deviates by 0.23 MM from the typical case, which does not exceed 12% of the period.

Table 2.2 Dependence of key DDL parameters on external factors and technological deviations

Name of parameter	Measurement unit	Minimum value	Typical value	Maximum value
Maximum delay step (MaxDS)	Unit range (UR)	0.09	0.12	0.13
Minimum delay step (MinDS)		0.029	0.03	0.06
Total delay interval of line (TLD)		2.06	2.27	2.5

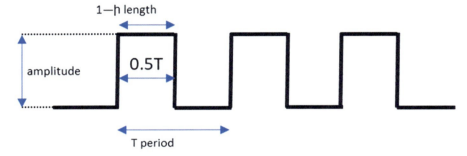

Fig. 2.24 Signal view with 50% duty cycle

2.1.3 Existing Problems of Design Methods of Integrated Circuits, Working Under Non-standard Operating Conditions

Existing means to design ICs that operate in non-standard operating conditions improve the accuracy of circuits, providing a certain stability against external factors. However, the development of the technological process, the reduction of signal-to-noise ratio, the increase in the effect of aging phenomena, as well as the sharp change of external conditions during IC operation give rise to challenges and difficulties, which are impossible to solve with existing methods in some cases. Below are the shortcomings and problems of the existing means of designing ICs that operate in non-standard conditions.

Existing Problems of Offset Reduction Method in Comparators

The main difficulty in using the offset reduction method in comparators is ensuring the symmetry of the switched used. It is necessary that the clock signals used for switching have a 50% duty cycle (Fig. 2.24); otherwise, it will lead to a difference in the times when the switches are on and off. Such a deviation of the clock signal can lead to incomplete discharge or charging of capacitors, which will mean there is a

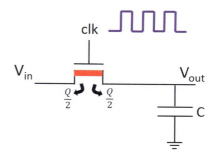

Fig. 2.25 In the ideal case, the dissipation of the accumulated charge in the channel

residual offset at the input of the comparator. In case of mismatch of several clock signals used in the system, the accuracy of the method decreases.

Taking into account the operation of circuits in non-standard operating conditions and sharp changes in temperature and voltage, ensuring such a value of duty cycle is a great challenge and requires a lot of effort for designers [88].

In addition to the mentioned drawback, there is another problem, which is due to the dissipation of the accumulated charge in the transistor channel when the switches are off. Since MOS transistors are used as a switch, the value of charge, diffused to the source and drain of the transistor when they are off, will depend on the resistance values of these pieces. If the resistances are equal, the accumulated charge will be divided and dispersed in two pieces. Taking into account the non-idealities of the technological process, it is not possible to ensure the uniform dispersion of the charge. The charge diffused into the potential will cause a change in the actual offset of the system stored in it (Fig. 2.25).

The amount of charge accumulated in the channel can be controlled by changing the physical dimensions of the transistor (2.6). Reducing the length and width of the gate will reduce its capacitance and reduce the amount of accumulated charge.

$$Q_r = (C_{ox} WLV_G \cdot V_{th}) \tag{2.6}$$

From the expression (2.7) it follows that it is not possible to reset the amount of accumulated charge to zero. On the other hand, reducing the size of the transistor will reduce the current passing through it and increase the charging and discharging times of the capacitor. This can cause the system to slow down. There are methods for reducing the residual offset caused by switch asymmetry, but their implementation requires increasing the circuit area and adding additional clock signals, while not completely solving the offset problem. Using capacitors in input cascade leads to an increase in the size of layout and can affect comparator bandwidth.

Considering the above-mentioned shortcomings, there is a need to develop new methods that will reduce the offset in the comparators without increasing the size of the circuit and having a negative impact on the stability of the system.

The Existing Problems in Offset Cancellation Method Based on Digital Calibration Circuit, for Input Stage of a Receiver

The main purpose of the receiver is to restore the distorted data signal after the channel. After eliminating the unwanted effect of the channel, only the signal is transmitted to the remaining blocks of the receiver. The presence of offset in the equalizer can lead to incorrect processing of the received signal, resulting in possible data loss. The operation of modern systems in non-standard conditions implies drastic temperature changes. The temperature dependence of transistor parameters can affect their state. Transition of the transistors from saturation mode to triode or off mode will cause the entire system to fail.

The simulation results show that in the method of offset reduction through the digital control circuit in IC receivers, the offset change as a result of temperature drift reaches up to 27 mV (Fig. 2.26d) (Table 2.3).

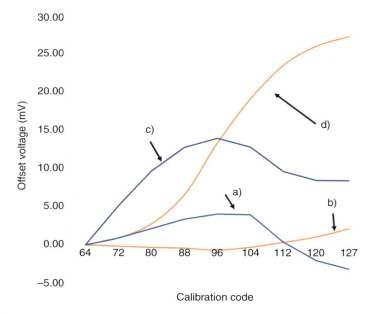

Fig. 2.26 Offset change

Table 2.3 The offset change of the equalizer in the worst case

	Temperature-independent current	Current obtained from a circuit with constant conduction
Maximum offset change	10 mV	27 mV

2.1 General Issues of Design Methods of Integrated Circuits, Working... 79

Fig. 2.27 Circuit, describing the operation of the system

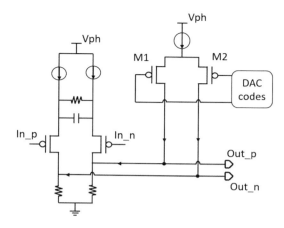

As mentioned above, the circuit can be powered by both temperature-independent and constant-conductance reference current sources. In the first case, the maximum offset of the system after calibration is 4 mV (Fig. 2.26a). In the second case, the maximum offset value is 2.2 mV (Fig. 2.26b).

In the first case, the offset change as a result of temperature drift is less than 10 mV (Fig. 2.26c), but in this case the system gain also depends on temperature.

The problem is that in the case of temperature-independent power supply and low calibration code, the circuit works as a differential amplifier (Fig. 2.27). Since the gain of the amplifier depends on the temperature, during the temperature drift there is a deviation of the state of the system from the calibrated state, therefore, also an offset change. In the case of high codes, only one of the M1 and M2 transistors works. Current flows through one branch. Since it is independent of temperature, the offset change of the system is less. But even then, the change in the state of the transistors causes the offset to increase.

When fed from a constant-conductance circuit, the offset change is larger for higher codes. Since the current flows through only one branch and has a strong dependence on the temperature, the offset value also changes when the latter fluctuates. At low codes, the differential pair works, and since it is fed from a constant-conductance circuit, the system conductance, and thus the gain, remains constant. In this case, the offset change is less. M1 and M2 transistors also cause a change in offset, because as a result of a change in temperature, there is a change in their threshold voltage and current in the saturation region.

The above described phenomena reduce the sensitivity of the system, reducing the efficiency of the calibration. Taking into account the shortcomings of the method, there is a need to develop new calibration mechanisms that will increase the quality characteristics of current ICs and ensure the minimum change of the offset value during the temperature drift.

Existing Problems in the Method of Continuous Reception of Sequential Data Using Digital Delay Lines

As it is known, as the temperature increases, the threshold voltage of the transistor decreases, which leads to an increase in the current in the saturation region (2.7).

$$I = \frac{1}{2}\mu C_{ox}\frac{W^*}{L}\left[(V_G\text{-}V_{th})^{2*}\left(1 + \frac{\Delta L}{L}\right)\right] \qquad (2.7)$$

where ΔL is the amount of channel shortening due to modulation.

An increase in the value of the supply voltage also leads to an increase in the current of the transistor in the saturation region, because the difference in the gate-source potentials increases (2.7). The speed of charging and discharging of the output load of the inverters and the input-output delay are directly related to the value of the current. In the case of small currents, the delay value is higher and vice versa. Therefore, varying the supply voltage and temperature individually or simultaneously can affect the total line delay value. Deviation of the signal by more than one unit range (UR) can result in missampling of the data signal. To check the effect of the above-mentioned phenomena on the operation of the circuit, after the DDL calibration phase, changes in temperature and supply voltage value were made. The temperature change was made gradually: $\pm 25\ °C$, $\pm 50\ °C$, and $\pm 100\ °C$ compared to the standard value of 25 °C. The value of the supply voltage has changed ± 30 mV compared to the value of 800 mV. Simulation results show a small change in the DDL step, at most 0,016 MM. Since all the inverters are affected in the same way by changes in external conditions, the step change between them is small. The total DDL delay, which is the sum of the delays of all inverters, has undergone a larger change. The maximum change is 0.61 MM. The difference between the maximum and minimum delays of the line increased from 0.44 MM to 1.08 MM, which is more than half of the signal period (Table 2.4).

Considering the increase in the frequency value in ICs, which has led to the reduction of the period of signals, similar changes in the line modulation can lead to an error in the reception of successive data.

The above changes caused by external conditions have a negative effect on the total delay cost of DDLs. Delay changes reduce the accuracy of calibration algorithm when operating the circuit under non-standard conditions. Taking all this into account, there is a need to develop new methods and measures for calibrating the DDL delay value, which will ensure the stability of the main parameters of the DDL after calibration, depending on the changes in external conditions. This will allow uninterrupted reception of sequential data in the case of circuit operation under non-standard conditions.

2.2 Reduction of Offset Value Caused by aging Phenomena Through...

Table 2.4 Change of main DDL parameters during temperature and voltage changes

Change in conditions	Measured value	Unit	Minimum value	Maximum value
±25 °C	MaxDS	UR	0.089	0.132
	MinDS		0.029	0.061
	TLD		2.05	2.52
±50 °C	MaxDS		0.087	0.133
	MinDS		0.028	0.063
	TLD		2.0	2.55
±100 °C	MaxDS		0.085	0.136
	MinDS		0.026	0.067
	TLD		1.91	2.59
+30 mV	MaxDS		0.088	0.128
	MinDS		0.027	0.058
	TLD		1.68	2.67
−30 mV	MaxDS		0.093	0.135
	MinDS		0.033	0.065
	TLD		1.66	2.74

2.1.4 Principles of Increasing the Stability of Integrated Circuits, Working Under Non-standard Operating Conditions

Taking into account the requirements for ICs, working in non-standard operating conditions and the unacceptable deviations caused by changes in temperature and supply voltage in existing methods due to aging of transistors and changes in threshold voltage, there is a need to develop new solutions and methods aimed at increasing the stability and efficiency of circuits. The following principles are suggested:

2.2 Reduction of Offset Value Caused by aging Phenomena Through Adding Transmission Gates and Transistor Switches in Comparators

Simulation results for one of the well-known structures of the described OpAmp showed that it has an input offset of up to 35 mV as a result of aging in the off state. The described method successfully reduced the offset value, but its implementation required a significant increase in area due to the addition of a second comparator cascade, the need for several clock signals for which a 50% duty cycle had to be provided, and capacitors that caused the circuit stability issues. It is recommended to keep the inputs of the circuit closed and ensure the minimum potential difference between the terminals of the remaining transistors due to the addition of transmission gates and MOS switches in the structure of the OpAmp, due to the minimum increase in area, because, as mentioned, in this case, the effect of aging phenomena is less.

2.3 Incorporation of the Current DAC in the Method of Offset Reduction Through the Circuit with Digital Control of IC Receiver

Taking the offset increase into account caused by the change in the parameters of the transistors feeding the outputs of the equalizer in the described method, it is recommended to eliminate them and use a current DAC instead of the DAC consisting of a circuit and resistors. Such a change, due to the reduction of the area occupied on a semiconductor crystal and the minimal increase in power consumption, allows the outputs of the circuit to be fed directly from the current DAC, the transistors of which have a greater saturation reserve and are stable during temperature changes.

2.4 Implementation of Negative Feedback in DDLs

Due to implementation of negative feedback, it is proposed to minimize the changes in the delays of the inverters used in the DDL circuit. One of the delay elements is used as a sensor in the new circuit to sense changes in external conditions. Its input voltages, which are also connected to DDL, are controlled by injecting additional current through feedback operation. The change in the threshold voltage of transistors, at the expense of neutralizing them with the voltage applied to the gate, minimizes the delay deviations (Fig. 2.28).

2.4.1 Conclusions

1. As a result of changes in external factors, the deviation of transistor parameters and the increase in the effect of aging phenomena have led to a decrease in the efficiency of the calibration nodes and equalizers included in integrated circuits, which can even lead to functional failures. Therefore, the development of design methods for integrated circuits that will work under non-standard operating conditions has become an extremely important challenge.
2. The existing means of designing integrated circuits working in non-standard operating conditions are based on calibration algorithms and are oriented to technological processes of larger sizes. In addition, the possibility of sudden

Fig. 2.28 Block diagram of the proposed method

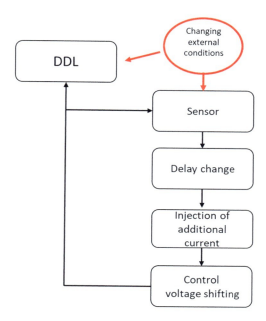

changes in temperature and voltage after calibration and the increased influence of aging phenomena typical of modern technologies are not taken into account. This indicates that the latter do not meet modern requirements, and the need to develop new solutions for increasing the stability and reliability of integrated circuits has arisen.

3. Approaches of IC design that work in non-standard operating conditions have been proposed, which will meet modern requirements and, at the expense of increasing the occupied area and power consumption within the permissible limits, will significantly reduce the deviations caused by changes in external conditions and aging phenomena.

2.5 Design Methods of Integrated Circuits, Working Under Non-standard Operating Conditions

The problems described in Sect. 2.1 witness about the shortcomings in IC design methods working under non-standard operating conditions. The increase in the involvement of ICs and the impact of aging phenomena in them and the sharp changes in voltage and temperature during the operation of devices have led to tightening the requirements towards IC reliability. Designers are forced to solve the problems that are present in the existing means or to develop methods to increase IC stability. Below are the proposed means of designing ICs, working under non-standard operating conditions, which solve the existing shortcomings in the existing methods, meeting the contemporary requirements.

2.5.1 Method of Reducing Offset, Caused by Aging Phenomena in Comparators

As mentioned above, one of the common comparator circuits has an offset problem caused by aging of the circuit in the off state. The proposed offset reduction method required the presence of capacitors, several clock signals, a second amplifier cascade, as well as switches. The problems related to IC stability, the complexity of ensuring the 50% duty cycle of the clock signal, and the dissipation of the charge when the switches are turned off caused a residual offset in the system. In order to avoid these defects, a method of offset reduction caused by aging phenomena in comparators was proposed [89].

Comparing the parameters of transistors in the on and off state of the comparator as a result of aging for 10 years, it becomes clear that the main reason for the occurrence of offset is the change in their threshold voltage. The maximum delta threshold was observed in the case of input transistors (Table 2.5) [89].

The effect of BTI and HCI phenomena on the input transistors is smaller when they are closed. Due to the addition of transmission gates TG1 and TG2, it is possible to cut off the inputs of the comparator and further control them in the off state of the circuit. D1 and D2 transistors have been added to control the inputs of the circuit. If they are open, the input transistors are in a closed state (Fig. 2.29) [89].

After the changes, the maximum offset of the comparator in the case of different input scenarios was 8 mV (Fig. 2.30).

The change in the threshold voltage of input transistors has decreased, reaching 131 ms. At the same time, there is still a change in the threshold voltage up to 191 mV in the circuit (Table 2.6) [89].

To reduce the changes in the threshold voltage of the current source M9 and M7 and M8 transistors, D3, D4, and D5 switches were added (Fig. 2.31) [89]. In the off state of the circuit, the D5 switch is open. The source terminals of M1 and M2 transistors, being connected to the supply voltage, provide a zero potential difference with respect to the gate. At the same time, the potential difference between all terminals of M9 transistor is zero, which leads to a reduction in the effect of aging phenomena. D3 and D4 switches ensure the discharge of outputs of M7 and M8 transistors to the ground when the circuit is off.

Table 2.5 The maximum delta threshold of the input transistors as a result of aging for 10 years for on and off states of the comparator

State of the circuit	The maximum delta threshold of input transistors (mV)
On	25.1
Off	273.2

2.5 Design Methods of Integrated Circuits, Working Under Non-standard... 85

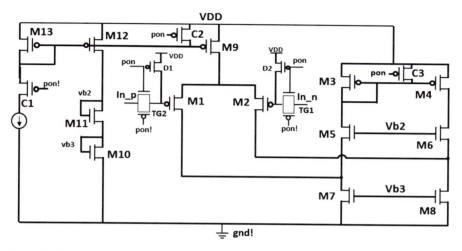

Fig. 2.29 Comparator schematic view after modifications

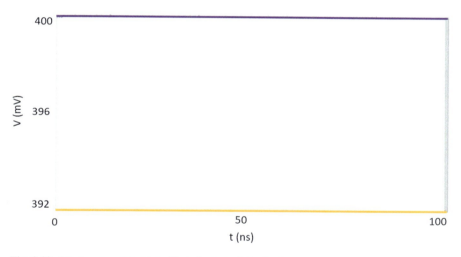

Fig. 2.30 Maximum comparator offset after transformations

Table 2.6 Maximum deviations of parameters of comparator transistors as a result of aging for 10 years

Name of transistor	Current shift from initial value (%)	Delta threshold (mV)
M9	9.8	191
M7/M8	8.03	155
M1/M2	6.32	131

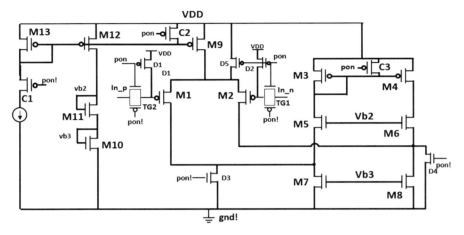

Fig. 2.31 Final view of the comparator

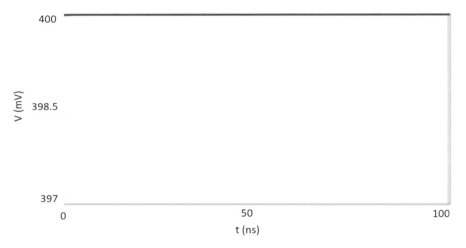

Fig. 2.32 The maximum offset of the final comparator circuit

After the specified modifications, the input offset value of the comparator is 3 mV (Fig. 2.32) [89].

In order to avoid leakage currents, as well as the described shortcomings, occurring when switching the transmission gates TG1 and TG2, the widths and lengths of the channel of the added transistor switches were chosen as small as possible.

Monte Carlo simulation was performed for typical offset and worst cases (Figs. 2.33 and 2.34) in order to check the effect of technological deviations on the operation of the circuit after modifications.

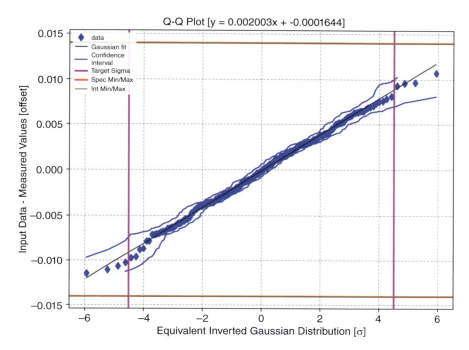

Fig. 2.33 Offset distribution as a result of Monte Carlo simulation for a typical case after transformations

The maximum offset value for typical and worst cases was 10.1 mV and 12.64 mV, respectively (Table 2.7).

A comparison of transistor parameters due to gain, offset, and aging was made with the results of the initial circuit under temperature conditions up to 150 °C (Table 2.8) [89].

Thus, at the expense of the 4.82% increase in the area of the described comparator, the proposed method ensures an 11.6 times reduction of the offset caused by the aging of the circuit. The area of the circuit has increased due to the addition of extra transistors and transmission gates. The gain of the comparator did not decrease as a result of modifications. The method does not require addition of extra outputs, since the transistors and transmission gates are controlled by the signals used to turn off the circuit.

2.5.2 Method of Reducing the Offset Caused by Sharp Fluctuations in Ambient Temperature in the Receiver

In the described method, the output offset of the receiver increased as a result of temperature fluctuations. The threshold voltage of M1 and M2 transistors was

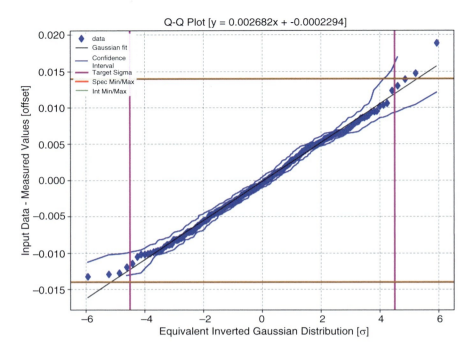

Fig. 2.34 Offset distribution as a result of Monte Carlo simulation for a worst case after transformations

Table 2.7 Results of a worst-case Monte Carlo simulation after transformations

Range	+4.5 sigma	−4.5 sigma
Offset (mV)	12.64	−11.63

Table 2.8 Comparison of parameters of the initial and final circuits of the comparator as a result of aging for 10 years

Parameter	Modified circuit	Initial circuit
Delta threshold (mV)	0.01	273
Offset (mV)	3	35
Gain (dB)	80.01	76.3
Maximum current shift (mkA)	0.1	9.8

changing. Both cases of choosing the current sources feeding the system had their drawbacks. A method is proposed [90] in order to avoid the changes of the offset.

The following changes have been made in the circuit:

1. To avoid voltage-current-voltage conversions, the pair of M1 and M2 transistors was added [90].
2. The structure of the DAC consisting of resistors was replaced by the current DAC (Fig. 2.7) [90] (Fig. 2.35).

2.5 Design Methods of Integrated Circuits, Working Under Non-standard...

Fig. 2.35 The current DAC circuit

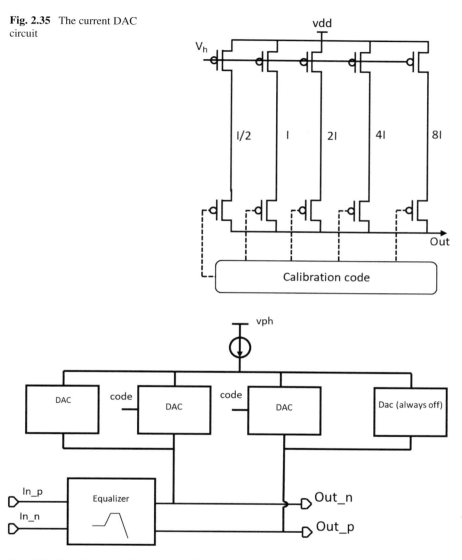

Fig. 2.36 The view of the DAC circuit after changes

A similar structure allows to connect current DAC directly to the outputs of the equalizer (Fig. 2.36) [90].

To simulate the case of getting voltages equal to the gates of M1 and M2 *transistors* and to keep the logic of calibration algorithm the same, an always-on current DAC was added. In order to ensure the symmetry of the branches in the layout, a disconnected DAC was added. In this way, it is possible to avoid the occurrence of additional offset in the output of the circuit. A similar structure allows

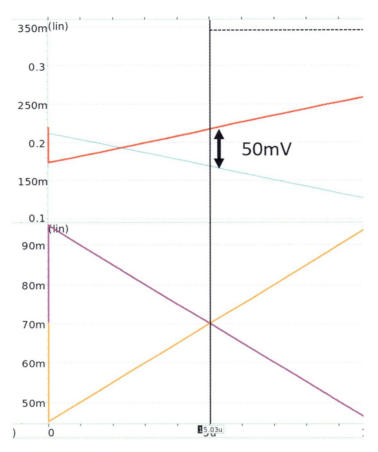

Fig. 2.37 Offset of an equalizer

the use of a temperature-independent reference current source for powering the circuit. It reduces the offset dependence on the current during sudden temperature changes.

In order to check the performance of the proposed method, one of the inputs was given an offset of 50 mV, and calibration algorithm was applied (Fig. 2.37) [90].

Before starting the calibration phase, the two inputs of the equalizer are brought close to each other from values of 50 and 100 mV. In the case of equality of inputs, the difference of the output branches will be exactly the offset. During calibration (Fig. 2.38) [90], the inputs of the equalizer are equalized to zero (Fig. 2.39). Changing the code leads to an increase in the value of the current given to branches. The two points of intersection of the outputs correspond to the minimum offset value. In the case of those points, the calibration code is fixed, after which the arithmetic mean is subtracted from them. The resulting outcome is considered the offset cancellation code corresponding to the given conditions.

2.5 Design Methods of Integrated Circuits, Working Under Non-standard... 91

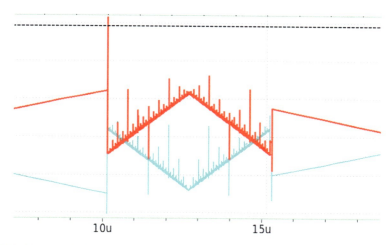

Fig. 2.38 Process of calibration

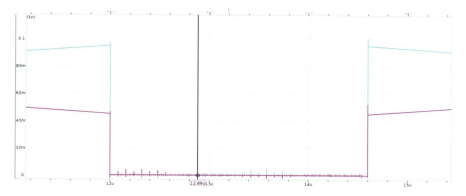

Fig. 2.39 Inputs of the cell during calibration

After deciding the code, the inputs of the node are brought together again. In this case, at the point of their intersection, the output offset already has a significantly smaller value (Fig. 2.40) [90].

As a result of the operation of calibration algorithm, the output offset was in decimal mV. The offset change due to the −40 to 150 °C thermal drift of the proposed method was 1.42 mV [90]. It was possible to achieve such a small change in the offset as a result of the application of the current DAC. The transistors used in it have a fairly large supply of saturation. Delta threshold as a result of the change in temperature does not cause a change in the state of transistors. In this approach the range of offset calibration is more linear, because the voltage is obtained directly on resistors of the equalizer (Fig. 2.41) [90].

To evaluate the effect of technological deviations on the proposed method, Monte Carlo verification was performed at 4.5 sigma range. Before using the method, the

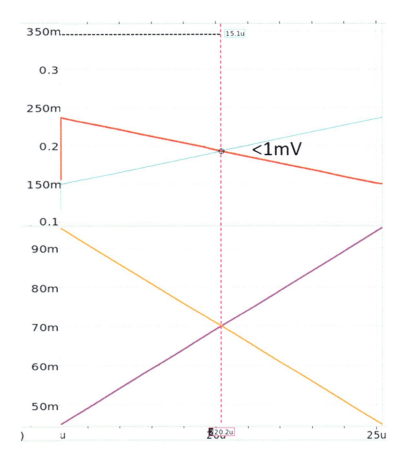

Fig. 2.40 Offset value after calibration

maximum value of offset was 4 mV in the range of ±4.5 sigma (Fig. 2.42) [90]. After the changes in the circuit, the maximum offset value in the worst and typical cases was 1.2 mV and 0.92 mV, respectively (Table 2.9) (Figs. 2.43 and 2.44) [90].

Thus, the replacement of DAC of resistors allowed to reduce the area of the circuit by 43.2% at the expense of refusing from transistor-resistor structure (Table 2.10) [90]. The current DAC occupies a smaller area, consisting only of MOS transistors. The power consumption increased by 7.2% due to the always-on power supply branch.

The obtained results prove that the proposed solution is effective in terms of reducing the output offset of the equalizer. Due to the reduction of the area and the increase of the power consumption, the method allows to achieve an offset of about 19 times the reduction in case of sharp changes in temperature.

2.5 Design Methods of Integrated Circuits, Working Under Non-standard... 93

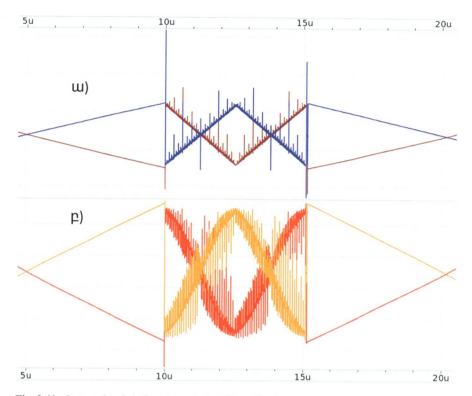

Fig. 2.41 Output signals before (**a**) and after (**b**) application of the method

2.5.3 Method of Reducing the Offset Caused by Sharp Voltage and Temperature Fluctuations in Digital Delay Lines

In the described method, the timing parameters of DDL deviated during temperature and voltage fluctuations. This was due to the change in the threshold and gate-source voltages of transistors in inverters.

This led to deviations in the charge and discharge times of output loads, which affected the DDL delay.

To avoid the above problem, a method based on negative feedback is proposed [91]. To sense temperature and voltage changes, an additional delay element was added to which feedback was applied (Fig. 2.17) [91]. In order to accurately measure changes in external conditions on the line, it is recommended to place the additional delay element in the physical design as close to the DDL as possible (Fig. 2.45).

The input and output of the added delay element are connected to the phase detector (PD). It represents an XOR circuit and is designed to capture delay deviation

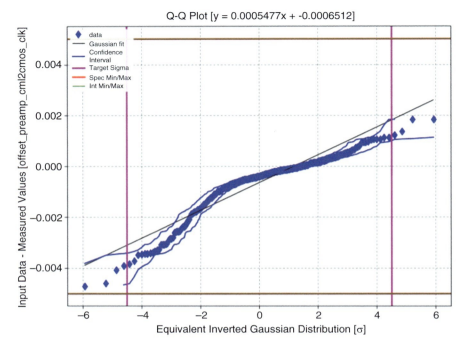

Fig. 2.42 The worst-case offset distribution before applying the method

Table 2.9 Results of a worst-case Monte Carlo simulation after modifications

Range	+4.5 sigma	−4.5 sigma
Offset (mV)	0.4	−1.2

due to changes in external conditions. The PD output is connected to a low-pass filter (LPF), which works as an integrator (Fig. 2.46) [91].

The LPF output is connected to a conductance OpAmp (COpAmp) (Fig. 2.47), which compensates for the change in delay by injecting additional current into the circuit that generates the delay controlling voltages (DCL) (Fig. 2.48) [91].

Depending on the external conditions, the number of appropriate mirrors is selected by the code in the COpAmp circuit. Then, depending on the number of mirrors, the total COpAmp current used to obtain the output voltages is determined. First, Vp and Vn voltages are given to the pMOS of the delay element, then to the gates of the nMOS transistors. By changing the gate-source voltage, the current flowing through the transistors is controlled. Depending on the value of the current, the delay of the circuit changes. The calibration process is considered complete.

After calibration, the threshold voltage of transistors changes during changes in temperature and voltage. This affects the value of the current flowing through the mirrors, leading to offset of Vp and Vn. Since the code is already fixed, the number of connected mirrors cannot be changed. By injecting additional current at the

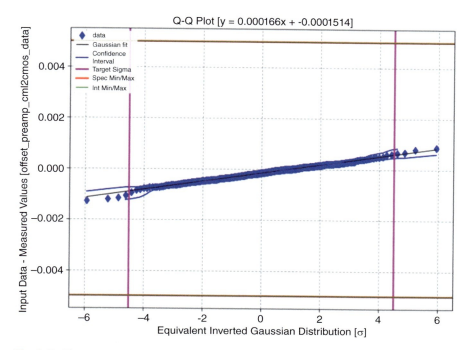

Fig. 2.43 The worst-case offset distribution as a result of using the method

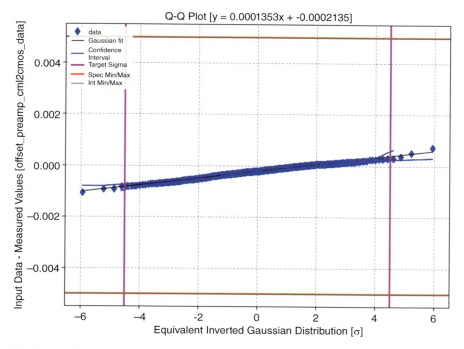

Fig. 2.44 Offset distribution in a typical case as a result of using the method

Table 2.10 Comparative results of the maximum offset change, the occupied area, and the power consumption for the circuits using the initial and proposed methods

Circuit	For the existing method	For the proposed method
Maximum delta threshold (mV)	26.98	0,0.42
Area (mkm^2)	2361.31	1341.2
Power consumption (mW)	8341.25	8941.82

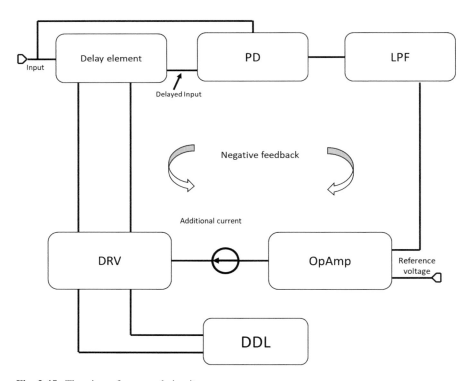

Fig. 2.45 The view of proposed circuit

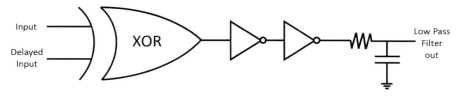

Fig. 2.46 XOR and LPF

2.5 Design Methods of Integrated Circuits, Working Under Non-standard... 97

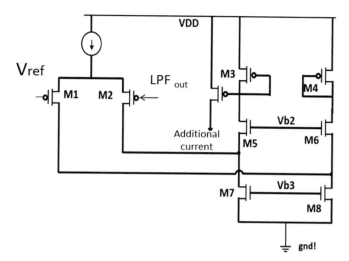

Fig. 2.47 COpAmp view

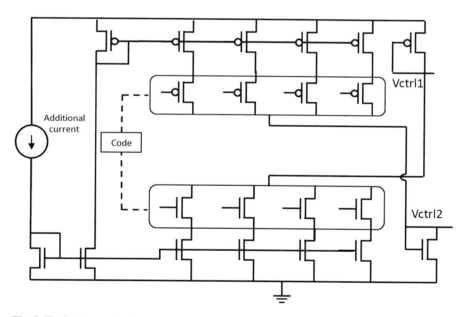

Fig. 2.48 COpAmp circuit

COpAmp output, the control voltages Vp and Vn are changed. Through feedback, these voltages are brought to values where the delay has minimal deviation. In addition to the feedback circuit, the voltages Vp and Vn are also connected to the DDL, since changes in external conditions have affected it in the same way (Fig. 2.49).

Fig. 2.49 DDL circuit

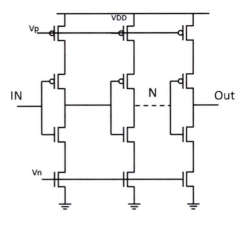

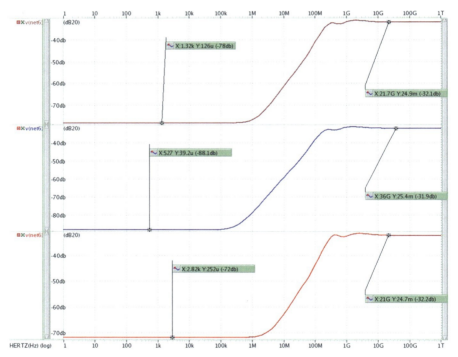

Fig. 2.50 COpAmp PSRRs for TT, FF, and SS cases

To check the reliability of COpAmp, its power supply rejection ratio (PSRR) (Fig. 2.50) [91] and gain (Fig. 2.51), amplification, and phase supplies were measured (Table 2.11). LPF cut-off frequency was chosen in the range of 120–180 kHz (Fig. 2.52).

2.5 Design Methods of Integrated Circuits, Working Under Non-standard...

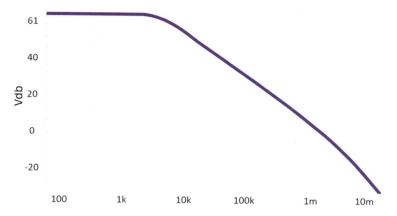

Fig. 2.51 COpAmp gain for worst case

Table 2.11 Simulation results for COpAmp

Parameter name	Worst-case results
Gain (dB)	61
PSRR (dB)	< -32
Phase margin (degree)	57
Gain margin (dB)	14.4

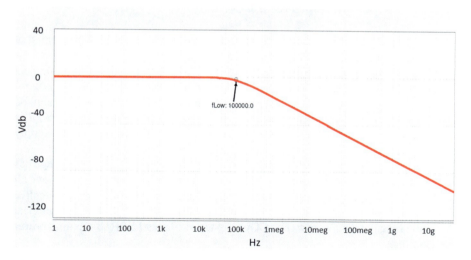

Fig. 2.52 Amplitude-frequency characteristic of LPF

PSRR is one of the most important parameters of COpAmp. Inadequate PSRR can cause gate-source voltage fluctuations of transistors, leading to a change in current value.

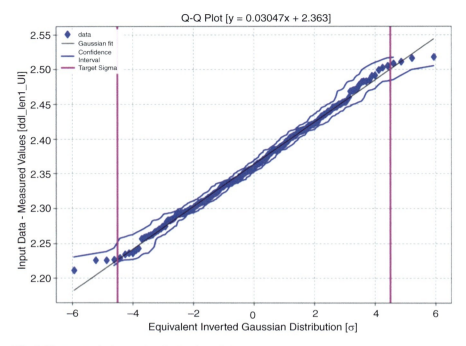

Fig. 2.53 In a typical case, the distribution of DDL delay

To verify the effectiveness of the method, Monte Carlo check was performed in the worst and typical cases [91] (Figs. 2.53 and 2.54).

The delay distribution is linear and satisfies the ±4.5 sigma range of technological deviations. After the changes made, the operation of the circuit was checked for changes in voltage and temperature. Changes in external conditions were applied after calibration was completed to verify the operation of negative feedback. The maximum range of DDL delay was 0.47 MM at −30 mV deviation of the supply voltage (Table 2.12) [91].

Thus, due to changes in temperature and supply voltage, the DDL delay range has been reduced by 56.04%, from 1.08 MM to 0.47 MM. It was possible to achieve all this due to the 23.1% increase in the area occupied on the die, because the negative feedback circuit was introduced. The circuit satisfies the Monte Carlo technological deviation range of ±4.5 sigma. The COpAmp has sufficient PSRR to suppress power jitter noises.

In summary, it can be said that the proposed method is effective in reducing deviations in DDLs operating under non-standard conditions.

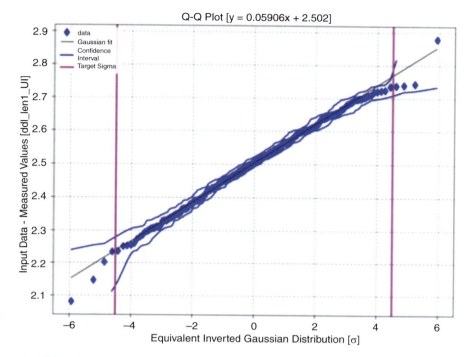

Fig. 2.54 In the worst case, the distribution of DDL delay

Table 2.12 Comparative DDL delay results for circuits using the initial circuit and proposed method

	DDL-h TLD (UR)			
	For initial circuit		For proposed method	
Change in external conditions	Min value	Max value	Min value	Max value
After calibration	2.06	2.5	2.28	2.58
±100 °C	1.91	2.59	2.2	2.69
+30 mV	1.68	2.67	2.17	2.63
−30 mV	1.66	2.74	2.11	2.58

2.6 Conclusions

1. Approaches of IC design, working in non-standard operating conditions, have been proposed, which meet modern requirements and, due to the increase of the occupied area and power consumption within the permissible limits, have significantly reduced the deviations caused by changes in external conditions and aging phenomena.

2. A method of reducing the offset caused by aging phenomena in comparators has been developed, in which, due to the addition of extra transmission gates and digital switches, the operating conditions of transistors in the off state of the circuit have been improved, and the offset has decreased by about 11 times, due to a maximum increase of 4.8% of the area.
3. A method of reducing the offset caused by sharp fluctuations in ambient temperature of IC receiver was designed, in which, due to the use of current DAC, it was possible to reduce the offset change of the equalizer by a maximum of 19 times, with a 43.2% decrease in the area occupied on the die and due to the 7.2% increase in power cost.
4. A method for reducing deviations caused by sharp voltage and temperature fluctuations in DDLs was created, in which, due to the introduction of negative feedback, it was possible to reduce the delay interval of DDL due to changes in external conditions by 56.04%—at the expense of 23.1% growth in area.

References

1. R. Schaller, Moore's law: Past, present and future. IEEE Spectr. **34**(6), 52–59 (1997). https://doi.org/10.1109/6.591665
2. T.P. Dash, S. Dey, E. Mohapatra, et al., Vertically-stacked silicon nanosheet field effect transistors at 3nm technology nodes. Devices for integrated circuit (DevIC) (2019), pp. 99–103, doi: https://doi.org/10.1109/DEVIC.2019.8783300
3. Z. Zhang, J. Sun, W. Zhu, et al., Design of a 3.2GHz~50GHz ultra wideband YIG-tunable-filter. International conference on microwave and millimeter wave technology (ICMMT) (2019), pp. 1–3, doi: https://doi.org/10.1109/ICMMT45702.2019.8992423
4. R. Yousry, 11.1 a 1.7pJ/b 112Gb/s XSR transceiver for intra-package communication in 7nm FinFET technology. IEEE international solid- state circuits conference (ISSCC) (2021), pp. 180–182, doi: https://doi.org/10.1109/ISSCC42613.2021.9365752
5. A. Cevrero, 6.1 a 100Gb/s 1.1pJ/b PAM-4 RX with dual-mode 1-Tap PAM-4/3-Tap NRZ speculative DFE in 14nm CMOS FinFET. IEEE international solid-state circuits conference (ISSCC) (2019), pp. 112–114, doi: https://doi.org/10.1109/ISSCC.2019.8662495
6. M. Rohith, K. Sreelakshmi, Design and integration of gateway electronic control unit (ECU) for automotive electronics applications. Asian conference on innovation in technology (ASIANCON) (2021), pp. 1–4, doi: https://doi.org/10.1109/ASIANCON51346.2021.9545049
7. V. Puranik, Sharmila, A. Ranjan, et al., Automation in agriculture and IoT. 4th international conference on internet of things: smart innovation and usages (IoT-SIU) (2019), pp. 1–6, doi: https://doi.org/10.1109/IoT-SIU.2019.8777619
8. J. Ouyang, X. Du, Y. Ma, et al., Kunlun: A 14nm high-performance AI processor for diversified workloads. IEEE international solid-state circuits conference (ISSCC) (2021), pp. 50–51, doi: https://doi.org/10.1109/ISSCC42613.2021.9366056
9. A. Yarali, Artificial intelligence, 5G, and IoT. Intelligent connectivity: AI, IoT, and 5G: IEEE (2022), pp. 251–268, doi: https://doi.org/10.1002/9781119685265.ch14
10. L. Wang, G. Liu, J. Wang, A method of temperature drift compensation for pulse synchronization in high-speed signal acquisition. 3rd IEEE international conference on control science and systems engineering (ICCSSE) (2017), pp. 529–534, doi: https://doi.org/10.1109/CCSSE.2017.8087988

References

11. S. Shin, L. Yongjae, J. Park, et al., A clock distribution scheme insensitive to supply voltage drift with self-adjustment of clock buffer delay. IEEE Trans. Circuits Syst. II Express Briefs **69**, 814–818 (2021). https://doi.org/10.1109/TCSII.2021.3110409
12. L. Luo, Y. Wu, J. Diao, et al., Low power low noise amplifier with DC offset correction at 1 V supply voltage for ultrasound imaging systems. IEEE 61st international midwest symposium on circuits and systems (MWSCAS) (2018), pp. 137–140, doi: https://doi.org/10.1109/MWSCAS.2018.8624065
13. B.T. Venkatesh Murthy, N.K. Singh, R. Jha, et al., Ultra low noise figure, low power consumption Ku- band LNA with high gain for space application. 5th international conference on communication and electronics systems (ICCES) (2020), pp. 80–83, doi: https://doi.org/10.1109/ICCES48766.2020.9137956
14. A. Boora, B.K. Thangarasu, K.S. Yeo, An ultra-low power 900 MHz intermediate frequency low noise amplifier for low-power RF receivers. IEEE 33rd international system-on-chip conference (SOCC) (2020), pp. 163–167, doi: https://doi.org/10.1109/SOCC49529.2020.9524753
15. P. Sandeep, P.A. Harsha Vardhini, V. Prakasam, SRAM utilization and power consumption analysis for low power applications. International conference on recent trends on electronics, information, communication & technology (RTEICT) (2020), pp. 227–231, doi: https://doi.org/10.1109/RTEICT49044.2020.9315558
16. K.-C. Chang, B.-Z. Lu, Y. Wang, et al., A 17.7-42.9-GHz low power low noise amplifier with 83% fractional bandwidth for radio astronomical receivers in 65-nm CMOS. IEEE Asia-Pacific microwave conference (APMC) (2020), pp. 507–509, doi: https://doi.org/10.1109/APMC47863.2020.9331381
17. V. Naranje, P.V. Reddy, B.K. Sharma, Optimization of factory layout design using simulation tool. IEEE 6th international conference on industrial engineering and applications (ICIEA) (2019), pp. 193–197, doi: https://doi.org/10.1109/IEA.2019.8715162
18. D.A. Bulakh, A.V. Korshunov, S.A. Ilin, Identification of integrated circuits based on layout layers routing information. IEEE conference of Russian young researchers in electrical and electronic engineering (ElConRus) (2021), pp. 1965–1968, doi: https://doi.org/10.1109/ElConRus51938.2021.9396208
19. A. Alshaabani, B. Wang, Parasitic capacitance cancellation technique by using mutual inductance and magnetic coupling. IECON 2019 – 45th annual conference of the IEEE industrial electronics society (2019), pp. 1928–1931, doi: https://doi.org/10.1109/IECON.2019.8927137
20. V.S. Melikyan, A.K. Mkhitaryan, H.T. Kostanyan, et al., Power supply noise rejection improvement method in modern VLSI design. IEEE East-West design & test symposium (EWDTS) (2019), pp. 1–4, doi: https://doi.org/10.1109/EWDTS.2019.8884372
21. D. Wu, C. Qian, X. Zhang, et al., Design of a capacitance measurement circuit with input parasitic capacitance elimination. IEEE 5th international conference on integrated circuits and microsystems (ICICM) (2020), pp. 53–57, doi: https://doi.org/10.1109/ICICM50929.2020.9292245
22. A. Khunteta, V. Niranjan, A novel noise reduction technique in CMOS amplifier. 3rd IEEE international conference on recent trends in electronics, information & communication technology (RTEICT) (2018), pp. 779–783, doi: https://doi.org/10.1109/RTEICT42901.2018.9012381
23. P.-C. Yeh, C.-N. Kuo, A 94 GHz 10.8 mW low-noise amplifier with inductive gain boosting in 40 nm digital CMOS technology. IEEE Asia-Pacific microwave conference (APMC) (2019), pp. 1357–1359, doi: https://doi.org/10.1109/APMC46564.2019.9038505
24. L. Fang, P. Gui, A 14nV/$\sqrt{Hz}$ 14μW chopper instrumentation amplifier with dynamic offset zeroing (DOZ) technique for ripple reduction. IEEE custom integrated circuits conference (CICC) (2019), pp. 1–4, doi: https://doi.org/10.1109/CICC.2019.8780239
25. K. Mikitchuk, A. Chizh, S. Malyshev, Noise and gain of an erbium-doped fiber amplifier for delay-line optoelectronic oscillator. International conference on noise and fluctuations (ICNF) (2017), pp. 1–4, doi: https://doi.org/10.1109/ICNF.2017.7985957

26. O. Bondarev, D. Mirvoda, A. Kosogor, et al., A line of 4–40 GHz GaAs low noise medium power amplifiers for SDH relay stations. 11th German microwave conference (GeMiC) (2018), pp. 187–190, doi: https://doi.org/10.23919/GEMIC.2018.8335061
27. Z. Wang, Z. Yuan, Y. Zhao, A gate-driver architecture with high common-mode noise immunity under extremely high dv/dt. IEEE applied power electronics conference and exposition (APEC) (2021), pp. 2532–2536, doi: https://doi.org/10.1109/APEC42165.2021.9487312
28. R. Raj, M.S. Bhat, S. Rekha, Library characterization: Noise and delay modeling. IEEE distributed computing, VLSI, electrical circuits and robotics (DISCOVER) (2018), pp. 44–48, doi: https://doi.org/10.1109/DISCOVER.2018.8674081
29. C. Prasad, S. Ramey, L. Jiang, Self-heating in advanced CMOS technologies. IEEE international reliability physics symposium (IRPS) (2017), pp. 6A-4.1–6A-4.7, doi: https://doi.org/10.1109/IRPS.2017.7936336
30. A.K. Mkhitaryan, G.A. Petrosyan, H.T. Grigoryan, et al., The reliability compensation method of voltage controlled oscillators. Proceedings of NPUA: Information technologies, electronics, radio engineering (2020), pp. 65–70
31. D. Son, G.-J. Kim, J. Kim, et al., Effect of high temperature on recovery of hot carrier degradation of scaled nMOSFETs in DRAM. IEEE international reliability physics symposium (IRPS) (2021), pp. 1–4, doi: https://doi.org/10.1109/IRPS46558.2021.9405153
32. R. Kishida, T. Asuke, J. Furuta, et al., Extracting BTI-induced degradation without temporal factors by using BTI-sensitive and BTI-insensitive ring oscillators. IEEE 32nd international conference on microelectronic test structures (ICMTS) (2019), pp. 24–27, doi: https://doi.org/10.1109/ICMTS.2019.8730967
33. S. Mishra, P. Weckx, J.Y. Lin, et al., Fast & accurate methodology for aging incorporation in circuits using adaptive waveform splitting (AWS). IEEE international reliability physics symposium (IRPS) (2020), pp. 1–5, doi: https://doi.org/10.1109/IRPS45951.2020.9129351
34. J. Diaz-Fortuny, J. Martin-Martinez, R. Rodriguez, et al., A noise and RTN-removal smart method for parameters extraction of CMOS aging compact models. Joint International EUROSOI workshop and international conference on ultimate integration on silicon (EUROSOI-ULIS) (2018), pp. 1–4, doi: https://doi.org/10.1109/ULIS.2018.8354740
35. A. Sivadasan, R. J. Shah, V. Huard, et al., NBTI aged cell rejuvenation with back biasing and resulting critical path reordering for digital circuits in 28nm FDSOI. Design, automation & test in Europe conference & exhibition (DATE) (2018), pp. 997–998, doi: https://doi.org/10.23919/DATE.2018.8342154
36. Y. Liu, X. Chen, Z. Zhao, et al., SiC MOSFET threshold-voltage instability under high temperature aging. 19th international conference on electronic packaging technology (ICEPT) (2018), pp. 347–350, doi: https://doi.org/10.1109/ICEPT.2018.8480781
37. X. Li, J. Qing, Y. Sun, et al., Linear and resolution adjusted on-chip aging detection of NBTI degradation. IEEE Trans. Device Mater. Reliab. **18**(3), 383–390 (2018). https://doi.org/10.1109/TDMR.2018.2847322
38. R.W. Johnson, J.L. Evans, P. Jacobsen, et al., The changing automotive environment: High-temperature electronics. IEEE Trans. Electron. Packag. Manuf. **27**(3), 164–176 (2004). https://doi.org/10.1109/TEPM.2004.843109
39. S.B. Yalçın,O. Demirci, M.E. Soltekin, Designing and implementing secure automotive network for autonomous cars. 29th signal processing and communications applications conference (SIU) (2021), pp. 1–4, doi: https://doi.org/10.1109/SIU53274.2021.9477958
40. I. Kastelan, M. Popovic, M. Vranješ, et al., Work in progress: Modernizing laboratories for innovative technologies in automotive. IEEE global engineering education conference (EDUCON) (2018), pp. 1700–1702, doi: https://doi.org/10.1109/EDUCON.2018.8363439
41. J. Zhou, X. Long, J. He, et al., Uncertainty quantification for junction temperature of automotive LED with die-attach layer microstructure. IEEE Trans. Device Mater. Reliab. **18**(1), 86–96 (2018). https://doi.org/10.1109/TDMR.2018.2796072

42. W.M.A. Halim, J.R. Rusli, S. Shafie, et al., Study on performance of capacitor-less LDO with different types of resistor. IEEE international circuits and systems symposium (ICSyS) (2019), pp. 1–5, doi: https://doi.org/10.1109/ICSyS47076.2019.8982395
43. P. Suriyavejwongs, E. Leelarasmee, W. Pora, A low voltage CMOS current comparator with offset compensation. IEEE asia pacific conference on circuits and systems (APCCAS) (2019), pp. 161–164, doi: https://doi.org/10.1109/APCCAS47518.2019.8953117
44. S. Pourashraf, J. Ramirez-Angulo, A.R. Cabrera-Galicia, et al., An amplified offset compensation scheme and its application in a track and hold circuit. IEEE Trans. Circuits Syst. II Express Briefs **65**(4), 416–420 (2018). https://doi.org/10.1109/TCSII.2017.2695162
45. L. Kouhalvandi, S. Aygün, G.G. Özdemir, et al., 10-bit high-speed CMOS comparator with offset cancellation technique. 5th IEEE workshop on advances in information, electronic and electrical engineering (AIEEE) (2017), pp. 1–4, doi: https://doi.org/10.1109/AIEEE.2017.8270524
46. Y.S. Vani, N.U. Rani, R. Vaddi, A low poltage capacitor based current controlled sense aamplifier for input offset compensation. International SoC design conference (ISOCC) (2017), pp. 23–24, doi: https://doi.org/10.1109/ISOCC.2017.8368810
47. A. Bamigbade, V. Khadkikar, DC-offset rejection approaches for single-phase frequency-locked loop. IEEE international conference on power electronics, drives and energy systems (PEDES) (2020), pp. 1–5, doi: https://doi.org/10.1109/PEDES49360.2020.9379567
48. V.S. Raja, S. Kumaravel, Design of recycling folded cascode amplifier using potential distribution method. International conference on microelectronic devices, circuits and systems (ICMDCS) (2017), pp. 1–5, doi: https://doi.org/10.1109/ICMDCS.2017.8211570
49. Y. Feng, Q. Fan, H. Deng, et al., An Automatic comparator offset calibration for high-speed flash ADCs in FDSOI CMOS technology. IEEE 11th Latin American symposium on circuits & systems (LASCAS) (2020), pp. 1–4, doi: https://doi.org/10.1109/LASCAS45839.2020.9069018
50. M. Wu, M. Lai, F. Lv, et al., An adaptive equalizer for 56 Gb/s PAM4 SerDes. 6th international conference on integrated circuits and microsystems (ICICM) (2021), pp. 398–402, doi: https://doi.org/10.1109/ICICM54364.2021.9660321
51. Chapter 2. Dynamic Offset Cancellation Techniques for Operational Amplifiers., https://ocw.tudelft.nl/wp-content/uploads/Reader_ET8017_Electronic_Instrumentation__DOC_techniques.pdf
52. D. Park, J. Kim, A 7-GHz fast-lock 2-step TDC-based all-digital DLL for post-DDR4 SDRAMs. IEEE international symposium on circuits and systems (ISCAS)(2018), pp. 1–4, doi: https://doi.org/10.1109/ISCAS.2018.8351396
53. T. Kim, J. Kim, A 0.8-3.5 GHz shared TDC-based fast-lock all-digital DLL with a built-in DCC. IEEE international symposium on circuits and systems (ISCAS) (2021), pp. 1–4, doi: https://doi.org/10.1109/ISCAS51556.2021.9401335
54. Y. Wei, S. Huang A folded locking scheme for the long-range delay block in a wide-range DLL. 2018 international SoC design conference (ISOCC) (2018), pp. 90–91, doi: https://doi.org/10.1109/ISOCC.2018.8649933
55. Z. Liu, L. Lou, Z. Fang, et al., A DLL-based configurable multi-phase clock generator for true-time-delay wideband FMCW phased-array in 40nm CMOS. IEEE international symposium on circuits and systems (ISCAS) (2018), pp. 1–4, doi: https://doi.org/10.1109/ISCAS.2018.8351374
56. X. Zhong, A. Bermak, C. Tsui, A low-offset dynamic comparator with area-efficient and low-power offset cancellation. IFIP/IEEE international conference on very large scale integration (VLSI-SoC) (2017), pp. 1–6, doi: https://doi.org/10.1109/VLSI-SoC.2017.8203481
57. J.-K. Han, J.-W. Kim, S.-H. Choi, et al., Asymmetrical half-bridge converter with zero DC-offset current in transformer using new rectifier structure. International power electronics conference (IPEC-Niigata 2018 -ECCE Asia) (2018), pp. 4049–4053, doi: https://doi.org/10.23919/IPEC.2018.8507457

58. L. Long, Y. Li, X. Liu A zero offset reduction method for RTD-based thermal flow sensors. 2021 IEEE international instrumentation and measurement technology conference (I2MTC) (2021), pp. 1–4, doi: https://doi.org/10.1109/I2MTC50364.2021.9459961
59. V. Raghuveer, K. Balasubramanian, S. Sudhakar, A 2μV low offset, 130 dB high gain continuous auto zero operational amplifier. International conference on communication and signal processing (ICCSP) (2017), pp. 1715–1718, doi: https://doi.org/10.1109/ICCSP.2017.8286685
60. M. Moezzi, S.F. Mousavi, P. Ashtari, An area-efficient DC offset cancellation architecture for zero-IF DVB-H receivers. IEEE Microw. Wirel. Compon. Lett. **28**(9), 813–815 (2018). https://doi.org/10.1109/LMWC.2018.2854259
61. D. Zeng, H. Zhu, W. Feng, et al., A 24-30GHz asymmetric SPDT switch for 5G millimeter-wave front-end. IEEE Asia-Pacific microwave conference (APMC) (2020), pp. 773–775, doi: https://doi.org/10.1109/APMC47863.2020.9331540
62. M.J. Rosario, F. Le-Strat, P.-F. Alleaume, et al., Low cost LTCC filters for a 30GHz satellite system. 33rd European microwave conference proceedings (IEEE Cat. No.03EX723C) (2003), pp. 817–820, doi: https://doi.org/10.1109/EUMC.2003.177601
63. H. Ahn, A. Dong, A. Wong, et al., 56Gbps PAM4 SerDes link parameter optimization for improved post-FEC BER. IEEE 28th conference on electrical performance of electronic packaging and systems (EPEPS) (2019), pp. 1–3, doi: https://doi.org/10.1109/EPEPS47316.2019.193195
64. https://nptel.ac.in/content/storage2/courses/117101058/downloads/Lec-8.pdf
65. A. Lapidoth, G. Marti, Encoder-assisted communications over additive noise channels. IEEE Trans. Inf. Theory **66**(11), 6607–6616 (2020). https://doi.org/10.1109/TIT.2020.3012629
66. V.S. Melikyan, A.S. Sahakyan, K.H. Safaryan, et al., High accuracy equalization method for receiver active equalizer. East-West design & test symposium (EWDTS 2013) (2013), pp. 1–4, doi: https://doi.org/10.1109/EWDTS.2013.6673119
67. K. Zheng, Y. Frans, K. Chang, et al., A 56 Gb/s 6 mW 300 um2 inverter-based CTLE for short-reach PAM2 applications in 16 nm CMOS. IEEE custom integrated circuits conference (CICC) (2018), pp. 1–4, doi: https://doi.org/10.1109/CICC.2018.8357076
68. Y.-H Chen, D.-B. Lin, Combined optimization of FFE and CTLE equalizations by analysis of pulse response. IEEE international symposium on radio-frequency integration technology (RFIT) (2021), pp. 1–3, doi: https://doi.org/10.1109/RFIT52905.2021.9565234
69. M.A. Dolatsara, H. Yu, J.A. Hejase, et al., Invertible neural networks for inverse design of CTLE in high-speed channels. IEEE electrical design of advanced packaging and systems (EDAPS) (2020), pp. 1–3, doi: https://doi.org/10.1109/EDAPS50281.2020.9312919
70. A. Balachandran, Y. Chen, C.C. Boon, A 32-Gb/s 3.53-mW/Gb/s adaptive receiver AFE employing a hybrid CTLE, edge-DFE and merged data-DFE/CDR in 65-nm CMOS. IEEE Asia Pacific conference on circuits and systems (APCCAS) (2019), pp. 221–224, doi: https://doi.org/10.1109/APCCAS47518.2019.8953146
71. M. Wen, L. Ding, X. Wang, et al., A 50 Gb/s serial link receiver with inductive peaking CTLE and 1-tap loop-unrolled DFE in 22nm FDSOI CMOS. IEEE MTT-S international wireless symposium (IWS) (2020), pp. 1–3, doi: https://doi.org/10.1109/IWS49314.2020.9360200
72. D. Lee, Y.-H. Kim, et al., A 0.9-V 12-Gb/s two-FIR tap direct DFE with feedback-signal common-mode control. IEEE Trans. Very Large Scale Integr. Syst. **27**(3), 724–728 (2019). https://doi.org/10.1109/TVLSI.2018.2882606
73. Y. Itoh, W. Xiaole, S. Omokawa, An L-band SiGe HBT active differential equalizer with tunable positive/negative gain slopes using transistor-loaded RC-circuits. Asia-Pacific microwave conference (APMC) (2018), pp. 708–710, doi: https://doi.org/10.23919/APMC.2018.8617315
74. K. Chen, W.W. Kuo, A. Emami, A 60-Gb/s PAM4 wireline receiver with 2-tap direct decision feedback equalization employing track-and-regenerate slicers in 28-nm CMOS. IEEE custom integrated circuits conference (CICC) (2020), pp. 1–4, doi: https://doi.org/10.1109/CICC48029.2020.9075948

75. C.F. Huang, Y.L. Chao, A hybrid de-embedding technique of eye diagram measurement for high-speed digital interconnections. IEEE transactions on components, packaging and manufacturing technology (2014), pp. 892–895
76. S. Guo, L. Ding, J. Jin, A 16/32Gb/s NRZ/PAM4 receiver with dual-loop CDR and threshold voltage calibration. IEEE 13th international conference on ASIC (ASICON) (2019), pp. 1–4, doi: https://doi.org/10.1109/ASICON47005.2019.8983675
77. H. Won, K. Han, S. Lee, et al., An on-chip stochastic sigma-tracking eye-opening monitor for BER-optimal adaptive equalization. IEEE custom integrated circuits conference (CICC) (2015), pp. 1–4, doi: https://doi.org/10.1109/CICC.2015.7338374
78. Hakob T. Kostanyan, Harutyun T. Kostanyan, G.A. Petrosyan, et al., 5V wide supply voltage bandgap reference for automotive applications. IEEE 39th international conference on electronics and nanotechnology (ELNANO) (2019), pp. 229–232, doi: https://doi.org/10.1109/ELNANO.2019.8783600
79. A.K. Mkhitaryan, H.T. Kostanyan, H.T. Grigoryan, et al., Stability improvement method for ultra-low-power bandgap reference. IEEE 40th international conference on electronics and nanotechnology (ELNANO) (2020), pp. 331–334, doi: https://doi.org/10.1109/ELNANO50318.2020.9088904
80. H. Seol, S. Hong, O. Kwon, An area-efficient high-resolution resistor-string DAC with reverse ordering scheme for active matrix flat-panel display data driver ICs. J. Disp. Technol. **12**(8), 828–834 (2016). https://doi.org/10.1109/JDT.2016.2526042
81. H. Zhu, Q. Sun, X. Li, Simulation of 112Gbps full-link interconnect system. Cross strait quad-regional radio science and wireless technology conference (CSQRWC) (2019), pp. 1–3, doi: https://doi.org/10.1109/CSQRWC.2019.8799127
82. T.-H. Tsai, R.-B. Sheen, C.-H. Chang, et al., A hybrid-PLL (ADPLL/charge-pump PLL) using phase realignment with 0.6-us settling, 0.619-ps integrated jitter, and −240.5-dB FoM in 7-nm FinFET. IEEE Solid-State Circuits Lett. **3**, 174–177 (2020). https://doi.org/10.1109/LSSC.2020.3010278
83. Z. Ge, J. Fu, P. Wang, Low power clock tree optimization by clock buffer/inverter reduction. IEEE international conference on integrated circuits, technologies and applications (ICTA) (2019), pp. 69–70, doi: https://doi.org/10.1109/ICTA48799.2019.9012842
84. C. Lin, S. Huang, W. Cheng, An effective approach for building low-power general activity-driven clock trees. International SoC design conference (ISOCC) (2018), pp. 13–14, doi: https://doi.org/10.1109/ISOCC.2018.8649800
85. M.F. Allam, A.A. Bdelrahman, H. Omran, et al., Novel decimation topology with improved jitter performance for clock and data recovery systems. 19th IEEE international new circuits and systems conference (NEWCAS) (2021), pp. 1–4, doi: https://doi.org/10.1109/NEWCAS50681.2021.9462785
86. V.G. Srivatsa, A.P. Chavan, D. Mourya, Design of low power & high performance multi source h-tree clock distribution network. IEEE VLSI device circuit and system (VLSI DCS) (2020), pp. 468–473, doi: https://doi.org/10.1109/VLSIDCS47293.2020.9179954
87. H. Wenjia, Y. Horii, Enhanced group delay of microstrip-line-based dispersive delay lines with lc resonant circuits for real-time analog signal processing. IEEE Asia Pacific microwave conference (APMC) (2017), pp. 272–275, doi: https://doi.org/10.1109/APMC.2017.8251431
88. Z. Zhang, W. Chu, S. Huang, The Ping-Pong tunable delay line in a super-resilient delay-locked loop. 56th ACM/IEEE design automation conference (DAC) (2019), pp. 1–2
89. H.T. Kostanyan, H.V. Margaryan, V.A. Janpoladov, et al., The minimizaton method of transistor ageing influence on modern voltage references. IEEE 40th international conference on electronics and nanotechnology (ELNANO) (2020), pp. 335–338, doi: https://doi.org/10.1109/ELNANO50318.2020.9088844
90. H.T. Kostanyan, The minimization method of thermal drift influence on analog integrated circuits. Proc. RA NAS NPUA. Series Tech. Sci. **75**(1), 120–128 (2022)
91. H.T. Kostanyan, Skew improvement method for digital delay lines operating in nonstandard environments. Proc. Univ. Electron. **27**(2), 233–239 (2022). https://doi.org/10.24151/1561-5405-2022-27-2-233-239

Chapter 3
Signal Transmission Calibration Systems in Integrated Circuits

3.1 General Issues of Signal Transmission Calibration Systems in Integrated Circuits

3.1.1 Importance of Signal Transmission Calibration Means in Integrated Circuits

I/O [3] cells are one of the most important components of contemporary ICs [1] (Fig. 3.1) [2]. They ensure lossless reading and transmission of data and also process the transmitted signal.

It should be noted that I/O cells can generally work in bidirectional mode, which means that the same I/O cell can both read the data and transmit it (Fig. 3.2).

Currently, there are many types of I/O cells [4], which have different applications, as a result of which the power supply voltage of I/O cells and the data transfer rates can be different. For example, double data rate (DDR) [5] nodes provide data transfer between the computer core and RAM [6]. There is also low-power DDR (LPDDR) [7], which is used in devices that require high power efficiency, such as mobile phones, modern laptops, and tablet computers. Another type of I/O cell is the Universal Serial Bus (USB) [8], which is used to transfer data and provide power between computers. Currently, high-bandwidth memories (HBM) [9] are also used, which provide data transfer between computer core and three-dimensional synchronous dynamic random access memory [10] (3SDRAM). Multimedia special I/O cells are also used [11, 12], for example, high-definition multimedia interface, which allows to transmit high-quality video data and audio data (Table 3.1) [13].

From Table 3.1 it can be noticed that the new generation I/O cells are faster and work with lower supply voltages, which significantly complicates the processes of data reading and transfer. It should be noted that as a result of reducing the supply voltage, ICs become sensitive to external and internal noise [14], and due to the increase in speed, the transmission line begins to significantly suppress the transmitted data, which makes data processing even more difficult.

110   3 Signal Transmission Calibration Systems in Integrated Circuits

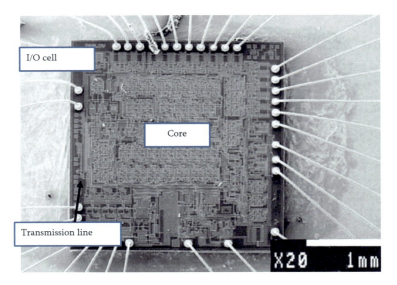

Fig. 3.1 IC structure

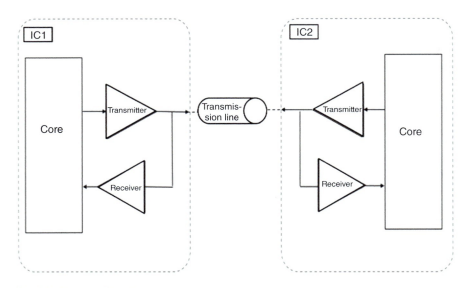

Fig. 3.2 Classical I/O cell structure

3.1 General Issues of Signal Transmission Calibration Systems...

Table 3.1 Subtypes of I/O cell

I/O cell standard	Data transfer speed (Gbps)	Supply voltage (V)
DDR4	1.6–3.2	1.2–1.6
DDR5	4.8–6.4	1.1–1.4
LPDDR4	3.2–4.2	0.9–1.1
LPDDR5	4.2–6.4	0.6–1.1
USB2	3–6	0.8–1.8
USB3	2.78–16	0.8–1.2
HBM	6–15	1.1–3.3
3SDRAM	5–16	0.9–1.8

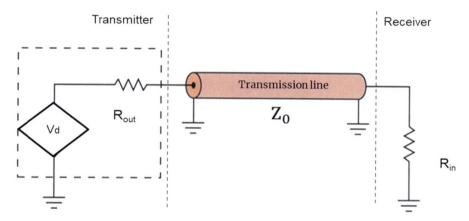

Fig. 3.3 Transmission line structure: V_d the transmitted signal, R_{out} the output of the transmitter, Z_0 the wave of the transmission line, and R_{in} the input resistance of the receiver

Currently, there are also high-precision IC types [15], which provide extremely high reliability. They are mainly used in different fields. Since the loss of data in the above areas can lead to great damage, I/O cells in high-precision ICs are equipped with systems to increase the reliability of data transmission and reading [16].

In ICs, I/O cells are connected via transmission lines [17], which contain capacitive and inductive components [18]. They significantly suppress the signal amplitude and cause distortions (Fig. 3.3) [19].

It should be noted that the wave resistance value of the transmission line is usually equal to 50 Ω [20].

Due to the suppressive properties of transmission lines and wave reflections, it is difficult to further increase the speed of ICs [21]. These reflections occur when the resistances of the transmitter and the receiver are not matched, which significantly reduces the reliability of the transmitted data, leading to even data loss (Fig. 3.4).

Depending on the degree of matching of transmitter and receiver resistances, the wave reflection coefficients will be different. For the general case, this coefficient is determined by the following formula [22]:

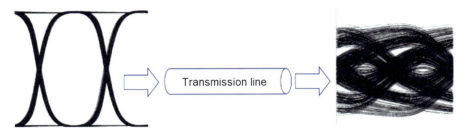

Fig. 3.4 Signal distortion in transmission lines

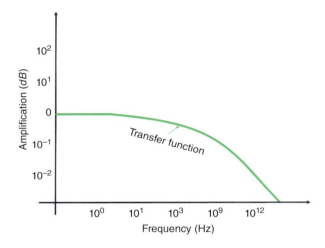

Fig. 3.5 Transmission line transfer function

$$\rho = \frac{R - Z_0}{R + Z_0}. \tag{3.1}$$

Signal reflections can occur both at the output of the transmitter and at the input of the receiver and are determined as follows [22]:

$$\rho_t = \frac{R_{out} - Z_0}{R_{out} + Z_0}, \tag{3.2}$$

$$\rho_\mathcal{L} = \frac{R_{in} - Z_0}{R_{in} + Z_0} \tag{3.3}$$

In modern I/O cells, data transfer rates can reach Gbps (Fig. 3.5) [23], which significantly complicates data reading. In that case, prevention of transmission line distortions becomes an even more important issue. For example, in fifth-generation double data rate nodes [24] (DDR5), the data transfer rate is 6.4 Gbps.

Reading a signal of such view (Fig. 3.6) requires special methods, the use of which will neutralize these distortions and allow lossless data reading [25–27].

3.1 General Issues of Signal Transmission Calibration Systems... 113

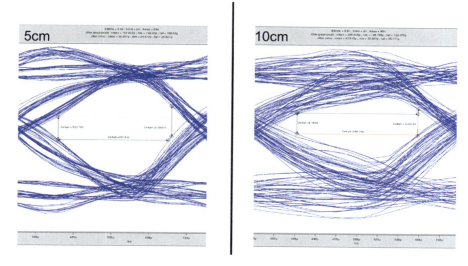

Fig. 3.6 Signal suppression property of transmission line

Developing accurate methods of data transfer and reading that will be applicable to a large number of I/O cells and will also neutralize the distortion caused by the transmission line is an important issue for companies that fabricate contemporary ICs.

In summary, the factors that hinder the process of increasing the speed of I/O cells and accurate data transmission, which causes a great demand for the development of means of regulating the transmission of signals, are:

- The large set of I/O cells and various operating requirements towards them
- Negative effect of the transmission line on signal transmission
- Increasing the performance of I/O cells
- High reliability of signal processing required for high-precision ICs.

3.1.2 Importance of Developing Signal Transmission Calibration Means in Integrated Circuits

In modern ICs, signal transmission calibration systems are mainly composed of I/O cells, the sub-nodes of which aim to increase the quality of data transmission and reading. The main components of the mentioned I/O cells are the transmitter and the receiver (Fig. 3.7) [28].

The data transmitted from the core is preprocessed in digital logic (DL) node. The latter implements the supply of control signals, determines the operating mode of I/O

114 3 Signal Transmission Calibration Systems in Integrated Circuits

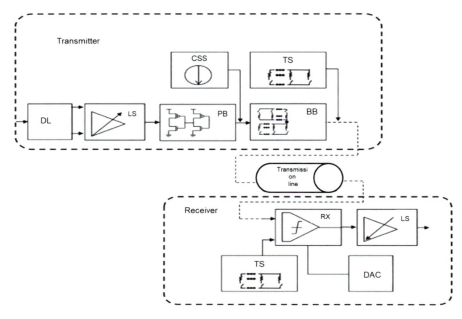

Fig. 3.7 Structure of I/O cell

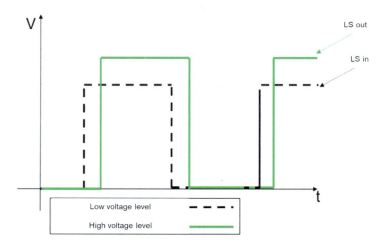

Fig. 3.8 Input and output signals of a level shifter

cell, as well as forms the signals characterizing the state of system's supply voltage and input signal.

Since the transmitted data and the core have different supply voltage values, a level shifter is used (Fig. 3.8) [29]. It should be noted that there is also a high-to-low level shifter, used in the receiver.

3.1 General Issues of Signal Transmission Calibration Systems...

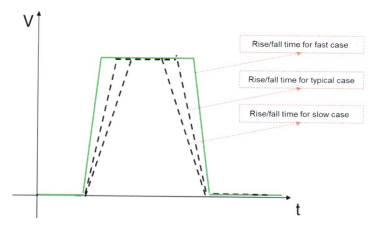

Fig. 3.9 Control of the speed of rise/fall times of transmitted data

The data, passing through the input node, is transferred to the output buffer (OB) [30], which is connected to the transmission line. The OB contains high-power buffers connected to each other in parallel, through which data is transmitted as lossless as possible. Depending on the operating mode of I/O cell, it may be required to provide different values of the output resistance of the transmitter. In that case, the DL node forms calibrating signals, as a result of which some of the OB nodes are connected or disconnected and change the value of the output resistance. Since the OB has a large input capacitance, the transmitted data is pre-amplified by a pre-buffer (PB) [31]. It consists of buffers that are connected to the current support system (CSS) [32]. The latter supplies additional current to buffers, as a result of which it is possible to control the speeds of rise/fall times of the transmitted data (Fig. 3.9), which are also controlled by DL node.

CSS can provide different values of speed of rise/fall times of the output signal. It depends on the type of I/O cell as well as the length of the transmission line. All these nodes aim to increase the reliability of the transmitted data as much as possible to reduce the impact of the transmission line. The latter has two types of models: for lossless and lossy transmission lines. The wave impedance for a lossless line is determined by the following formula [33]:

$$Z_0 = \sqrt{\frac{L}{C}}, \qquad (3.4)$$

and for the line with loss it is determined by the following formula [33]:

$$Z_0 = \sqrt{\frac{R + j\omega L}{G + j\omega C}}, \qquad (3.5)$$

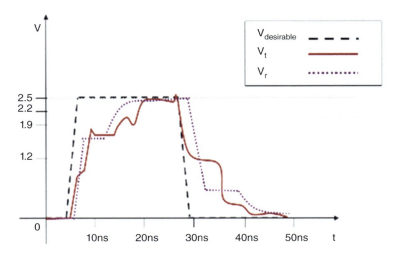

Fig. 3.10 The view of data distorted by wave reflection

where R is the resistance of the transmission line, G is the conductance of the transmission line, and $j\omega L$ and $j\omega C$ are inductive and capacitive components of the transmission line.

To avoid signal distortion (Fig. 3.10), it is extremely important that the output resistance of the transmitter, the wave impedance, and the input resistance of the receiver are coordinated [34, 35].

$$R_{out} = Z_0 = R_{in}, \qquad (3.6)$$

However, impedance matching becomes difficult in the case of process-voltage-temperature (PVT) deviations [36], as a result of which the transmitter's output and receiver's input impedance values deviate, causing wave reflections and causing transmitted data error. Fig. 3.11 shows the effect of temperature change on the resistance value.

In order to neutralize the influence of PVT deviations, termination system (TS) [37] is used, which allows to keep the transmitter output and receiver input resistance values unchanged. The TS compares the value of the resistance outside the IC with high accuracy (±2% accuracy) with the output resistance, which changes during the PVT variations [38–40]. If deviations are detected, some of the component buffers of the OB are connected on or off in parallel with the OB of the transmitter, as a result of which the value of the output resistance of the OB decreases or increases. By performing this process, it is possible to reach the target resistance value and avoid wave reflections as a result of its mismatch (Fig. 3.12). It should be noted that the TS is also used to match the input resistance in the receiver. Thus, the signal, passing through the transmission line, begins to be read at the data receiver (DR) [41], which is composed of high-frequency receivers [42, 43]. Depending on

3.1 General Issues of Signal Transmission Calibration Systems... 117

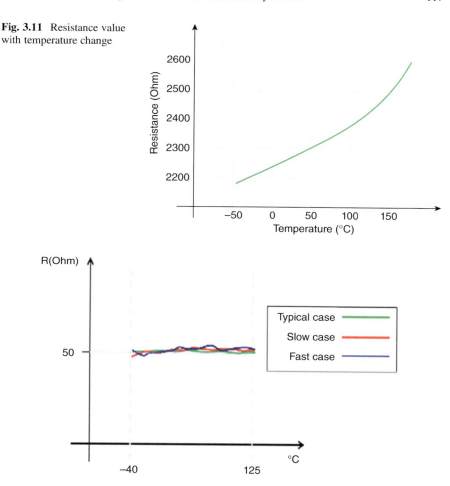

Fig. 3.11 Resistance value with temperature change

Fig. 3.12 The neutralization of resistance fluctuations by means of TS

the type of I/O cells, it may consist of cascades of several receivers. It aims to amplify the data at its target frequency and suppress other frequencies as much as possible, which will restore the data distorted by the transmission line.

These receivers usually work in reference voltage comparison or differential modes. In reference voltage comparison mode, the reference voltage is applied to the negative input of the receiver, and the transmitted data is applied to the other input. That reference voltage is supplied from a digital-to-analog converter (DAC) in the receiver [44, 45]. In this mode, the receiver presents a comparator and, in cases of values higher than the reference voltage, generates logic "1" at the output, and in cases of values lower than the reference voltage, logical "0" signals (Fig. 3.13). It should be noted that ideally, the value of the reference voltage should be equal to half

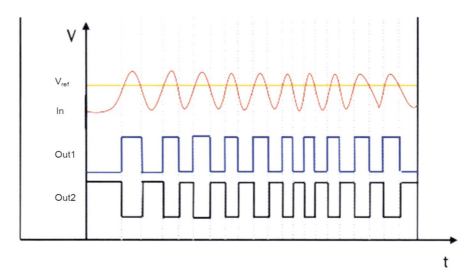

Fig. 3.13 Receiver's reference voltage in comparison mode

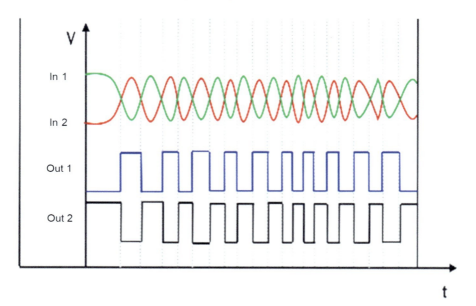

Fig. 3.14 Receiver operation in differential mode

the amplitude of the transmitted data, so that the signal is transmitted to the core as evenly as possible.

In the differential mode, signals deviated 180 degrees are applied to the receiver inputs (Fig. 3.14) [46]. In this mode, the receiver is more noise-immune, because noises contained in the input data are not transmitted to the output.

3.1 General Issues of Signal Transmission Calibration Systems... 119

Problems of Developing Signal Transmission Calibration Means in Integrated Circuits and Their Measurements

With the increase of data transfer rates, deviations of timing parameters of data become an extremely important problem [47]. In modern I/O cells, the fastest data change time reaches ps [48]. In that case, even small deviations of timing parameters can lead to data error. The most common way of measuring timing parameters of a signal is the eye diagram (Fig. 3.15) [49], through which the size of the signal jitter, voltage levels, as well as the main timing parameters are observed. The rectangle depicted in the diagram is called the eye opening, the horizontal side of which corresponds to the time margin of the data, and the vertical side to the voltage corresponds to the amplitude of the data.

Another important timing parameter is the signal duty cycle which is determined by the following formula [50]:

$$D = \frac{T_1}{T_p} * 100\%, \qquad (3.7)$$

where D is the duty cycle of signal, T_1 the duration of the logical "1" range of the signal, and T_p the period of the signal.

For an ideal case, the signal duty cycle should be equal to 50%. It means that the durations of logic "1" and logic "0" of the signal are equal, and the signal has a proportionally periodic view. Mainly, duty cycle deviations are caused by the asymmetry of rise/falls of the signal [51], which in turn is caused by transmission line effects [52] and PVT deviations [53].

Another important timing parameter of the data is jitter (Fig. 3.16) [54–57]. It can be represented as the sum of signal deviations, reflections, data-dependent interference [58], noises affecting the data, and delays. It is generally calculated for a given unit period. The common jitter consists of two main components: deterministic [59] and arbitrary jitters (Fig. 3.17) [60].

The occurrence of deterministic jitter is not accidental; it has clear reasons and is usually periodic. It means that the jitter can be repeated at one or more frequencies. It consists of two subgroups: data-dependent and periodic jitters. Symbolic

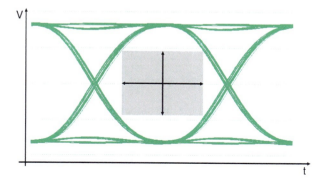

Fig. 3.15 Eye diagram

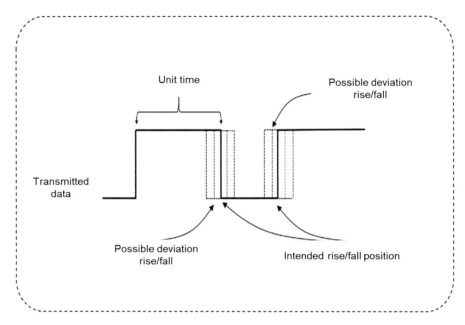

Fig. 3.16 The occurrence of data jitter

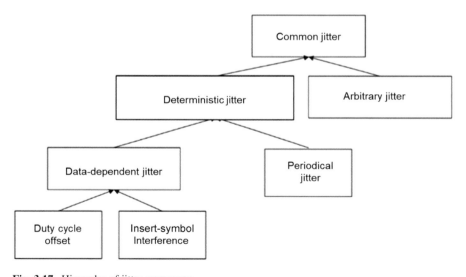

Fig. 3.17 Hierarchy of jitter occurrence

interference [61] is a type of data-dependent jitter observed during data transmission. For example, when transmitting sequential data "00001" and ""11101," their last switching jitters may be different. It is caused by reflections and frequency losses of the transmission line (Fig. 3.18). The latter can be described by the following formula [62]:

3.1 General Issues of Signal Transmission Calibration Systems...

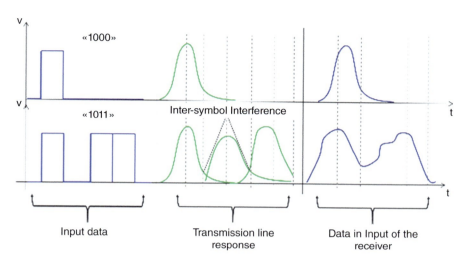

Fig. 3.18 Occurrence of inter-symbol interference

$$y(t) = \sum_{n=-\infty}^{\infty} \chi[n] * h(t - nT_s), \qquad (3.8)$$

where $y(t)$ is the signal formed at the receiver input, $h(t)$ is the impulse response of the transmission line, T_s is signal period, and $\chi[n]$ is the data being sent.

It can be observed that the response of the transmission line is different for the "0001" and "1011" data, so at the data "1011" a distorted signal is formed at the input of the receiver. Such distortions can lead to data loss, resulting in system crashes. Therefore, it is extremely important to ensure lossless data transfer.

Thus, one of the most relevant problems in current I/O cells is the development of means of signal transmission calibration, which will enable to significantly improve the reliability of data transmission, as well as to reduce the negative effects of the transmission line.

3.1.3 Current State and Issues of Design Methods of Signal Transmission Calibration Systems in Integrated Circuits

The demand for the design of signal transmission calibration systems [63] in ICs is conditioned by difficulties of data transmission and reading caused by the increase in speeds in I/O cells [64].

In addition, modern ICs are more sensitive to PVT deviations [65]. Under these conditions, accurate data processing is extremely difficult and can often cause system failures, which is a critical issue in high-precision ICs that operate in highly sensitive environments and require high data processing accuracy. The development of signal transmission calibration means will allow not only to significantly increase the reliability of data transmitted in the high frequency range, but also to significantly reduce the design time.

The increase in the reliability of transmitted data is conditioned by the use of signal transmission calibration means [66], due to which it will be possible to neutralize the negative effect of the transmission line, and they will be more stable to wave reflections, and it will also be possible to neutralize signal distortions caused by PVT deviations [66–70].

The reduction of design time is conditioned by the increase in data reliability as a result of the use of signal transmission calibration means [71–74]. Currently, the design of I/O cells is a rather long-lasting process, because the design is carried out with consideration of all possible deviations, as well as testing I/O cells under the conditions of possible PVT deviations [75, 76]. It should be noted that in this case it is often not possible to ensure the required performance of I/O cells in all cases.

Signal Reading Problems Caused by Wave Reflections at the Receiver

As it was mentioned, one of the most important timing parameters of reading signal is its duty cycle [77]. In an ideal case, it should be equal to 50%, but due to wave reflections, noises, and PVT deviation, it can deviate, causing data reading error [78]. To solve this problem, wide-range duty cycle calibration system (Fig. 3.19) [79] is used in receivers.

The latter calibrates the deviations of duty cycle by controlling the current of NMOS and PMOS transistors connected to the input buffer. These transistors are controlled by the control signal $V_{controlling}$ formed at the output of the operational amplifier. Depending on the value of this voltage, the current of NMOS and PMOS transistors connected to the input buffer increases or decreases, as a result of which it is possible to control the duty cycle of the input signal. The indicated control voltage $V_{controlling}$ is formed from the difference in the values of the voltages applied to the inputs of the differential amplifier. A voltage equal to half of the supply voltage is applied to the positive input of the differential amplifier, which corresponds to a signal with a duty cycle of 50%, and the output of the RC filter is connected to the other input [80]. The RC filter integrates the input data, as a result of which the resulting voltage value corresponds to the duty cycle of the signal.

For example, in the result of integrating a signal with an amplitude of 1 V and a duty cycle of 50%, 0.5 V is obtained, and a signal with a duty cycle of 60% corresponds to 0.6 V. Thus, by applying these two voltages to the inputs of the operational amplifier, it is possible to detect the direction of the deviation of duty cycle and correct it as a result of current support. However, this method has low accuracy, and at high frequencies, it is not possible to ensure accurate calibration of duty cycle.

3.1 General Issues of Signal Transmission Calibration Systems...

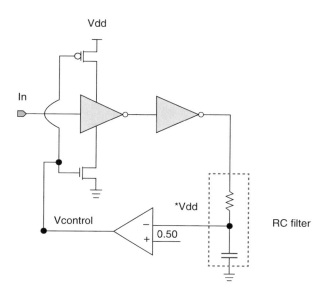

Fig. 3.19 Wide-range duty cycle calibration system

Problems Caused by Data Distortions at the Transmitter Sub-node

As a result of increase in data transfer rate, the input capacitances of sub-blocks in I/O cells begin to significantly suppress the transmitted data. A key part of that capacitance is the gate-source capacitance of a MOS transistor. In particular, high-speed data transmission in level shifters (LP) becomes difficult using the classical structure. For this reason, high-frequency LPs are used (Fig. 3.20) [81], in which the input node is the output of a MOS transistor [82], in order to reduce the effect of its gate-source capacitance.

High-frequency LSs with such a structure are used in almost all modern I/O cells, but the disadvantage of this structure is that it distorts the data. As a result of LP being the drain of the input transistor, the rise/fall of the transmitted data at the interconnection points "A" and "B" are distorted (Fig. 3.21). The distorted data of this form is transmitted to the output, as a result of which the signal's duty cycle is violated, as well as large jitter of data occurs [83, 84].

These deviations affect the time margin of the transmitted data and in some cases even cause data losses. Therefore, it is obvious that such a structure of LS limits the further increase of data transfer rate.

Problems Caused by PVT Deviations in Transmitter Sub-nodes

Modern I/O cells must provide full performance under various PVT conditions. However, PVT deviations cause current, output resistance, and voltage deviations [85, 86], which negatively affects the transmitted data and can even cause system

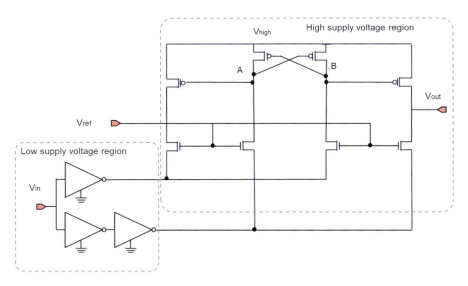

Fig. 3.20 High-frequency level shifter

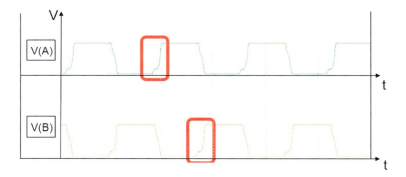

Fig. 3.21 Data distortion in high-frequency LS

error. For this reason, the current support system (CSS) is used (Fig. 3.22) [87], which provides additional current support and corrects the deviated rise/falls of the transmitted data.

CSS contains imitation of output buffer that is connected in parallel to the main output buffer. It can amplify rise/falls by applying "PMOS amplification" and "NMOS amplification" signals (Fig. 3.23).

Thus, it is possible to recalculate the deviations of the transmitted data, but the disadvantage of this system is that it does not perform rise/fall equalization during data transmission, as a result of which IC reliability is significantly reduced.

3.1 General Issues of Signal Transmission Calibration Systems... 125

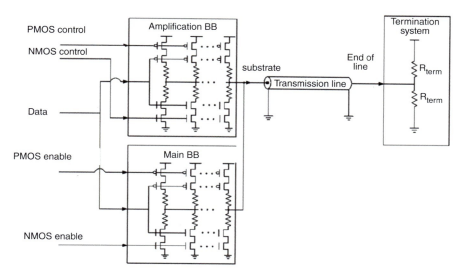

Fig. 3.22 CSS structure

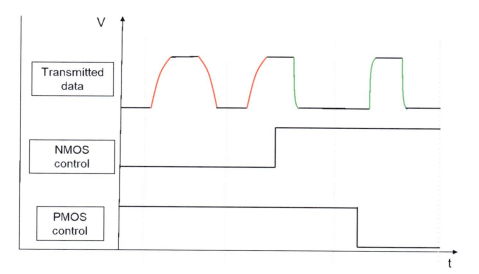

Fig. 3.23 Fixing the offset of the data passed through CSS

Signal Transmission Calibration Problems Caused by Transmission Lines

The increase in data speeds significantly complicates the process of their transmission. In particular, the transmission line's property to suppress the signal can cause

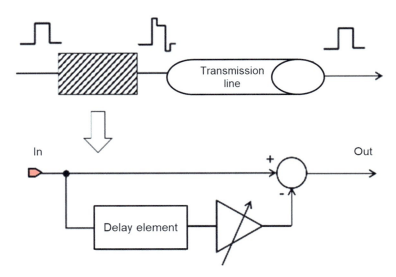

Fig. 3.24 Signal equalization method in transmitters

data loss, which limits further increases in data transfer rates. For this reason, signal equalization methods are used [88–90], which partially neutralize the negative effect of the transmission line. In general, the essence of the equalization method is similar to the operation of the finite impulse response filter; the input signal, passing through the delay element, is superimposed on the original input signal. As a result, a new type of signal is formed, which is amplified in high frequency ranges (Fig. 3.24) [91].

In modern I/O cells, the equalization method of signal amplitude modulation (SAM) is widely used, which is implemented by means of an "XOR" logic cell. Input and clock generated width modulation signals are applied to its inputs, as a result of which an already equalized signal is formed (Fig. 3.25) [91]. Moreover, depending on the value of duty cycle of the clock SAM, it is possible to reduce the inter-symbol interference of the signal.

Depending on the parameters of the transmission line, it is also possible to choose such a value of duty cycle of clock SAM, in the case of which the inter-symbol interference of the transmitted signal will be minimal. The value of this duty cycle can be calculated using the function of signal period (T_S) and channel time constant (T_C) of the transmission line [91].

$$\text{Duty cycle} = \frac{T_c}{T_s} = \ln\left(\frac{1}{2} + \frac{1}{2}e^{\frac{T_s}{T_c}}\right)\frac{T_c}{T_s}. \tag{3.9}$$

The disadvantage of this equalization method is that it is very difficult to accurately calculate the duty cycle in different PVT conditions, which leads to the error of the transmitted data. It is obvious that these equalization methods are not applicable for high-precision ICs, because they do not ensure high reliability of the transmitted data.

3.1 General Issues of Signal Transmission Calibration Systems...

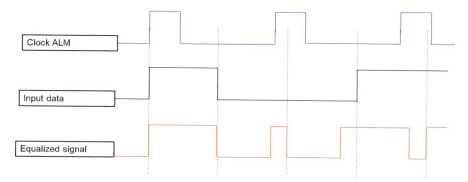

Fig. 3.25 Working principle of the SAM equalization method

Thus, the main obstacles of increasing the speed of data transfer in I/O cells and its lossless reading are PVT deviations, transmission lines, and inaccuracies caused by transmitters. In order to eliminate them, it is necessary to design signal transmission calibration systems. The research of the existing approaches and means of developing their design means shows that they do not meet the modern requirements for practical design from the point of view of efficiency.

Design Principles of Signal Transmission Calibration Systems in Integrated Circuits

It follows from the above written that currently used signal transmission calibration systems do not take power consumption problems into account, and they do not provide the required reliability of data transmission and reading, so they are not applicable in high-precision ICs and limit the further increase in the speed of I/O cells.

In particular:

- The wide-range duty cycle calibration system used in the receiver has low accuracy. Its use in case of high signal frequencies is not possible to ensure the exact calibration of duty cycle which can cause an error in the data that are read.
- The high-frequency level shifters used in the transmitter can distort the transmitted data, which complicates the accurate transmission of data. These deviations affect the time margin of the transmitted data and in some cases even cause data losses. Therefore, it is obvious that such a structure of level shifters limits further increase of data transfer rate.
- The current support systems used in the transmitter do not perform rise/fall calibration during signal transmission, as a result of which the reliability of integrated circuits is significantly reduced.
- It is difficult to calibrate the deviations due to wave reflections using signal width modulation equalization methods, which leads to the error of the transmitted data.

It is obvious that these equalization systems are not applicable to high-precision integrated circuits, because they do not provide high reliability of the transmitted data.

Thus, taking into account the existing problems, the following principles are proposed for the design of signal transmission calibration means:

- To design a new duty cycle calibration system, which will allow to implement calibration of duty cycle during signal transmission, as a result of which it will be possible to ensure high reliability of the read data.
- To implement a new high-frequency level shifter architecture in the transmitter, which will not distort the transmitted signal and will ensure high reliability of the transmitted data.
- To reduce the dependence on process-voltage-temperature deviations in the current support node through the current control system and to implement the equalization of signal rise/falls during the operation of integrated circuits through its new structure, as a result of which the reliability of the transmitted data will increase significantly.
- To implement such a method of equalization system that will allow to reduce the effect of the transmission line and increase the reliability of reading the transmitted data.

Thus, principles of developing signal transmission calibration means have been proposed, which will allow to significantly improve their main technical characteristics and parameters: speed, reliability of data transmission, and reading.

Conclusions

1. One of the most relevant problems in the creation of modern integrated circuits is the development of signal transmission calibration means, which can significantly improve their main parameters and contribute to further increase in the speed of I/O cells.
2. The main obstacles to increasing the speed of data transmitted in I/O cells and their lossless transmission are process-voltage-temperature deviations, the influence of transmission lines, and deviations caused in transmitter sub-connections. Development of signal transmission calibration means is an effective way to eliminate them.
3. Research of existing approaches and means of signal transmission systems in integrated circuits shows that although they provide a significant improvement in speed, data transmission and reading reliability, and power consumption, from the point of view of efficiency, the latter do not meet the modern requirements for practical design.
4. Principles of developing signal transmission calibration means in integrated circuits have been proposed, which will allow to significantly improve their main technical characteristics and parameters: speed, reliability of data transfer and reading, and consumed power.

3.2 Design Principles of Signal Transmission Calibration Systems in Integrated Circuits

3.2.1 Method of Detection and Self-Regulation of Duty Cycle Deviations in High-Speed Integrated Circuits

As mentioned in Sect. 3.1, deviations in the signal duty cycle at the receiver can reduce the reliability of the read data, causing system failures. One of the main factors of their occurrence are PVT deviations, which can appear both before the transmission of the signal and during its transmission (Fig. 3.26). However, currently used duty cycle calibration systems are mostly applied before the signal transmission process. That is the reason that deviations during signal transmission are not taken into account, and the reliability of the system decreases significantly.

Thus, it is obvious that in order to increase the reliability of I/O cells, deviations during signal transmission should also be taken into account.

The proposed duty cycle calibration system (Fig. 3.27 [92]) detects and calibrates the duty cycle deviations during the entire operation of the I/O cell.

In modern I/O cells, the processing of the transmitted data is carried out by reading the data around the reference voltage. It is supplied either through the DAC in the receiver [93] or through the external DAC. The proposed calibration system, detecting deviations in the duty cycle of the signal, modifies the input code of the internal DAC of the receiver and leads to a change in the reference voltage. As a result, calibration of the duty cycle of the received signal occurs.

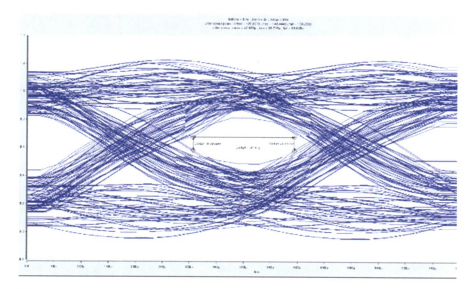

Fig. 3.26 Deviations that occur during signal transmission

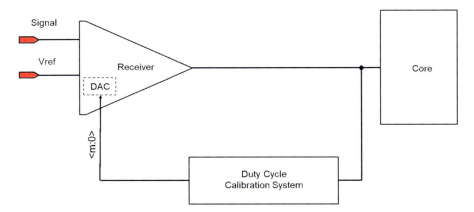

Fig. 3.27 Placement of duty cycle calibration system in I/O cell

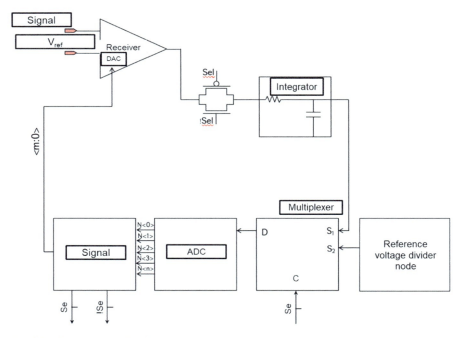

Fig. 3.28 Duty cycle calibration system structure

The main components of the designed calibration system (Fig. 3.28) are RC integrator, ADC, reference voltage division node (RVDN), analog multiplexer [94–96], as well as digital logic node.

3.2 Design Principles of Signal Transmission Calibration Systems... 131

Fig. 3.29 Structure of an RC integrator

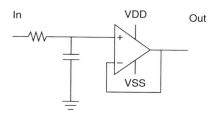

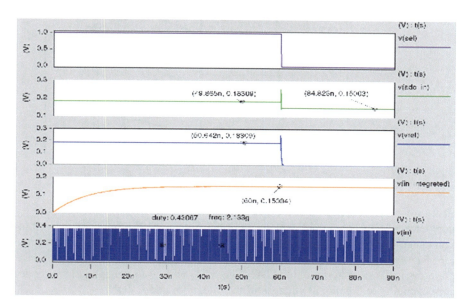

Fig. 3.30 Operation of an RC integrator

It should be noted that such a structure allows not to interrupt the data transfer process and carry out the calibration in parallel, neutralizing the duty cycle deviations caused by PVT deviations and other reasons.

The purpose of using the RC integrator (Fig. 3.29) is to obtain the constant voltage level of the received signal, which expresses the value of the duty cycle. It consists of a classic RC integrator circuit and an analog repeater. The analog repeater is designed on the basis of operational amplifier and aims to keep the value of the integrated voltage constant.

Thus, in this way, it is possible to obtain the value of the duty cycle (Fig. 3.30) and detect its deviations. For example, integration of a signal with an amplitude of 1 V and a duty cycle of 50% will result in a constant signal voltage level of 0.5 V, and in the case of a duty cycle of 40%—0.4 V.

It can be noticed that the integration process does not happen instantaneously. In order to establish a voltage corresponding to duty cycle at the output of the

Fig. 3.31 Structure of used ADC

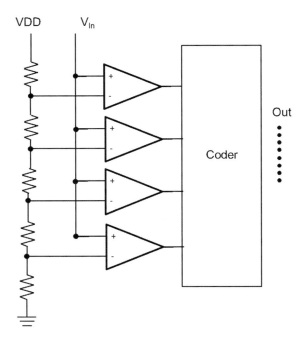

integrator, ~50 ns is required, which determines the duration of one system calibration cycle. The latter is calculated for the PVT in which the confirmation time is the longest.

In the calibration system, the ADC (Fig. 3.31) is used to generate digital codes corresponding to the output voltages of the integrator and RVDN.

It should be noted that the voltage provided by RVDN (Fig. 3.32) is equal to half of the amplitude of the received signal, that is, to 50% of the duty cycle. As a result of this process, digital codes are formed, one of which is equal to the value of the duty cycle at the given moment of the signal, and the other is equal to the value of 50% of duty cycle. Since there is a power-to-ground current path in the RVDN, a PMOS transistor is used to reduce the leakage current. It opens when the ADC is supplied with a voltage equivalent to 50% of the duty cycle. The supply of these voltages is carried out by means of an analog multiplexer, which selects one of the outputs of the RVDN or the integrator and connects it to the ADC.

A digital logic cell is used to detect deviations, which compares the output codes of the ADC. After the comparison operation, it supplies the DAC node with a new input code, as a result of which the reference voltage of the receiver changes and the duty cycle of the received signal is calibrated. This process continues until the duty cycle reaches the target 50%. It should be noted that the accuracy of the system can be improved by increasing the bitness of the ADC, which contributes to the detection of smaller deviations in the duty cycle.

Fig. 3.32. The structure of the reference voltage division node

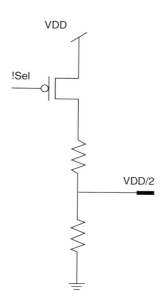

Estimation of Duty Cycle Deviation Detection and Efficiency of Self-Calibration Method

In order to evaluate the effectiveness of the proposed system, a high-frequency receiver was designed using a SAED 28 nm [97] technological process. Simulations were also performed under 27 PVT conditions, when the frequency of the received signal is 2133 MHz, which corresponds to the LPDDR4 standard. During simulations, the received signal previously contained ±15% duty cycle deviations, which the system calibrated (Fig. 3.33).

From the simulation results, it can be seen that the system calibrated the signal duty cycle to 50.1% within 3 μs. It can be noticed that a significant improvement of the duty cycle is already registered after 1 μs of the system operation. In particular, at 1.55 μs, the value of the duty cycle becomes 51.8%, and already at 2.25 μs, it becomes 50.9%.

The system digitizes reference and integrated voltage values at 25 ns intervals, since the signal integration process takes 50 ns for the slowest PVT condition. Therefore, the duration of one system calibration cycle is also equal to 50 ns. Thus, the sel signal updates the constant voltage level of the signal corresponding to duty cycle, and after the completion of each calibration cycle, it gets closer to the target value of the duty cycle (Fig. 3.34).

The simulation results (Table 3.2) show that the calibration system can work effectively in different PVT conditions. In the worst PVT condition, the system calibrated duty cycle, reaching it to 50.5%.

In general, it was possible to calibrate the pre-biased duty cycle signals for different PVT conditions and correct the duty cycle in the range of 49.7–50.5%.

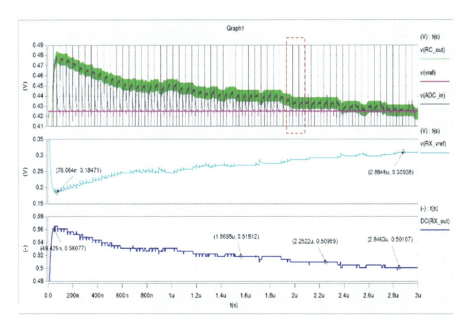

Fig. 3.33 Duty cycle calibration results under typical PVT conditions

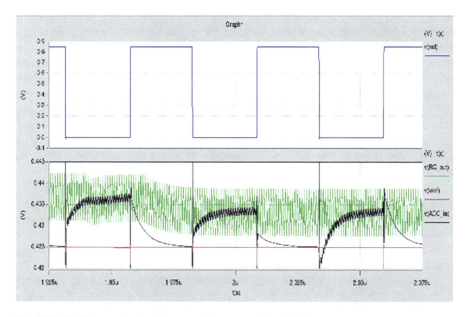

Fig. 3.34 ADC input selection between reference and signal DC voltages

3.2 Design Principles of Signal Transmission Calibration Systems...

Table 3.2 Simulation results of the proposed calibration system

PVT	Duty cycle	Power consumption (mW)	Supply voltage (V)
Typical, 25 °C	0.501	3.66	0.85
Fast, −40 °C	0.497	3.95	0.9
Slow, 125 °C	0.505	3.2	0.8
Slow, −40 °C	0.502	3.3	0.8
Fast, 125 °C	0.499	3.77	0.9

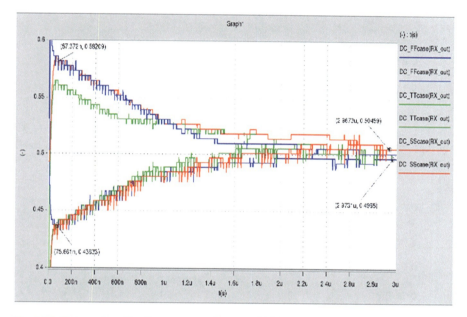

Fig. 3.35 Duty cycle calibration process under worst PVT conditions

Simulations showed that the system consumes a maximum power of 3.95 mW, which allows it to be used in low-power ICs. It should be noted that the proposed structure allows the calibration of the duty cycle to be carried out in the range of 40–60% deviations. That range was chosen taking into account the range of possible deviations of the duty cycle in modern ICs.

The results show that the duty cycle calibration time is almost the same for different PVT conditions and does not depend on the direction of deviations (Fig. 3.35).

The proposed mixed-signal system is applicable to both periodic signal and data sequences. For this purpose, a simulation was carried out for the PRBS5 data sequence with a frequency of 2133 MHz (Figs. 3.36 and 3.37).

While reading of the above data, the system changed the reference voltage of the receiver, as a result of which it became possible to restore the crossing point of the eye diagram.

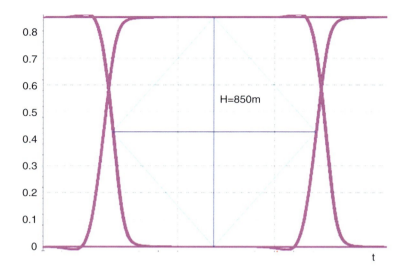

Fig. 3.36 Without using the calibration system, the eye diagram of the PRBS5 type data, being received

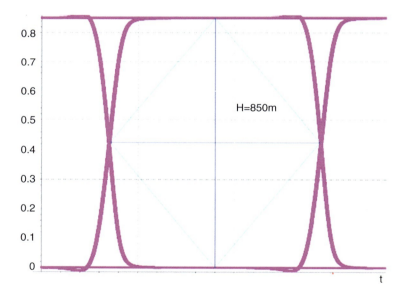

Fig. 3.37 Using the calibration system, the eye diagram of the PRBS5 type data, being received

A comparison with currently used duty cycle calibration systems was also carried out, which showed the effectiveness of the proposed system (Table 3.3).

Thus, as a result of using the proposed system, it is possible to significantly improve the reliability of the data during data reading and calibrate the duty cycle

3.2 Design Principles of Signal Transmission Calibration Systems...

Table 3.3 Simulation results of the proposed calibration system

Operation	[98]	[79]	[99]	[100]	Proposed system
Accuracy (%)	1	2	1	0.25	0.5
Calibration region (%)	30–70	25–75	30–70	20–80	40–60
Power consumption (mW)	20	1.4	1.1	3.2	3.95
Operating frequency (GHz)	3.5	2	0.6	1.7	2.1
Technological process (nm)	130	45	350	180	28
Supply voltage (V)	–	0.9–1.4	3.3	–	0.8–0.9

with an accuracy of ±0.5%. However, as a result of implementation of the system, the consumed power of the receiver will increase, which in the worst case is 3.95 mW.

3.2.2 Method of Increasing the Reliability of High-frequency Data in Transmitters

As it was mentioned, as a result of the increase in the data transfer rate, the input capacitances of sub-blocks in I/O cells begin to significantly suppress the transferred data. For this reason, in high-frequency LS (Fig. 3.38) [101], the output of a MOS transistor is used as an input to reduce the effect of its gate-source capacitance.

As a result of this structure, distortions of the transmitted data appear; in particular, step-like segments appear during the signal rise/fall. The latter significantly reduces the reliability of the transmitted data and negatively affects its time margin. Since LSs are used in almost all I/O cells, it is obvious that in order to increase the reliability of the data transmitted in the I/O cell, the signal distortion properties of high-frequency LSs should be neutralized.

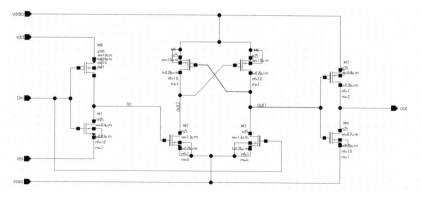

Fig. 3.38 Structure of high-frequency LS

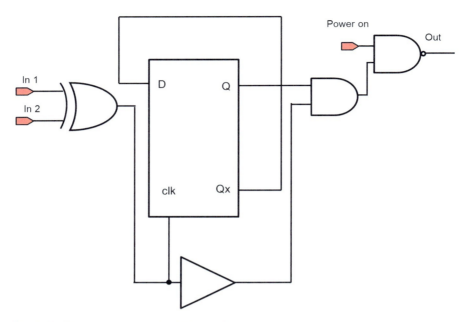

Fig. 3.39 The structure of the rise/fall correction system

Thus, a new high-frequency LS structure was designed, in which a rise/fall correction system was introduced (Fig. 3.39) [102], which allows to calibrate the signal jitter and duty cycle deviations, as well as reset the rise/fall target speeds.

The rise/fall correction system was implemented using digital logic cells. Its input node is the XOR, the transmitted data is applied to one of its inputs, and the delayed version of that data is applied to the other input. Thus, XOR creates pulse signals during the difference of inputs (Fig. 3.40) [103]. This signal informs about signal switches, during which the system should amplify the rise/fall times of the transmitted data. Since the system is implemented using digital logic cells, it occupies a small area.

For proper processing of the output signal of XOR, a frequency divider implemented on the basis of the DFF [103] was used, which is switched during rise times at the output of the XOR. As a result, pulse-shaped signals are formed at the output of the rise/fall correction node (Fig. 3.41).

The frequency divider supplies these signals to MOS transistors in high-frequency LS, which open during switching (Fig. 3.42) and provide additional current to the interconnection points "a" and "b."

The rise/fall correction node is connected in parallel to LS (Fig. 3.43), thus not interfering with the data transfer process.

3.2 Design Principles of Signal Transmission Calibration Systems...

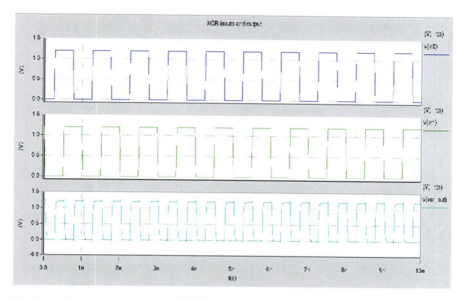

Fig. 3.40 Input and output signals of XOR

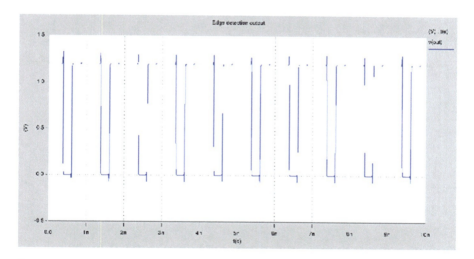

Fig. 3.41 Input and output signals of XOR

It should be noted that when the switching of the input signal occurs at one of the "a" or "b" interconnection points, P4 and P5 MOS transistors are in the open state (Fig. 3.44). In all other conditions, they are closed.

Thus, the proposed rise/fall correction system calibrates the distorted signals, increasing the reliability of transmitted data.

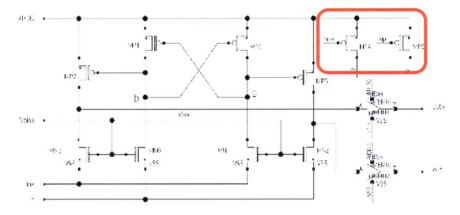

Fig. 3.42 The connection of additional MOS transistors in high-frequency LS

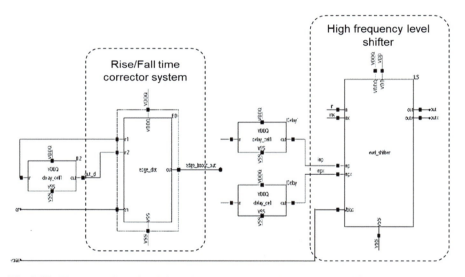

Fig. 3.43 The connection of additional MOS transistors in high-frequency LS

Evaluation of Effectiveness of the Method of Increasing the Reliability of High-frequency Data in Transmitter Sub-nodes

In order to evaluate the effectiveness of rise/fall correction system, a high-frequency LS was designed, as well as a signal correction node. Simulations were also performed at 27 PVT conditions for SAED 28 nm technology [97]. The measurements were carried out for two cases: when the rise/fall correction system is on and when it is off. The simulation results, without the use of the correction system, are presented in Table 3.4.

3.2 Design Principles of Signal Transmission Calibration Systems...

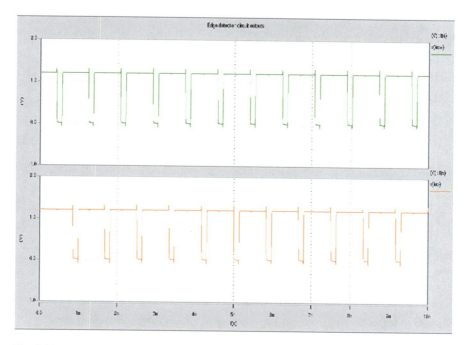

Fig. 3.44 Opening signals of P4 and P5 MOS transistors

Table 3.4 Simulation results without applying the correction system

Parameter	PVT				
	Typical, 1.2 V, 25 °C	Fast, 1.26 V, 125 °C	արագ, 1.26 V, −40 °C	Slow, 1.14 V, 125 °C	Slow, 1.14 V, −40 °C
Duty cycle (%)	49.5	49.3	51.1	48.7	49
Jitter (ps)	4.16	5.83	7.5	10.83	8.33

It can be observed that without the application of the correction system, the output signal of the high-frequency LS contains a large jitter (10.83 ps), which occurs at a slow PVT condition of 1.14 V, 125 °C (Fig. 3.45).

In the case of using the correction system, the step-like sections at the interconnecting points "a" and "b" disappear (Fig. 3.46).

The simulations showed that when using the system, the step-like sections disappear in all PVT conditions (Fig. 3.47).

The simulation results of high-frequency LS in the case of using the correction system are presented in Table 3.5.

It can be noticed that the jitter of the output signal is significantly reduced when using the system. The system also calibrated the duty cycle values; before using the system, the duty cycle values ranged between 48.7% and 51.1%, and after the system was applied, the range decreased significantly, reaching 49.55–50.5% (Fig. 3.48).

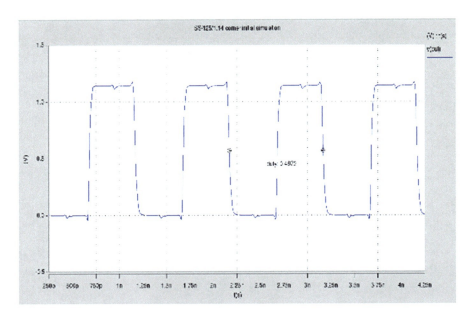

Fig. 3.45 View of the high-frequency LS output signal without using correction system

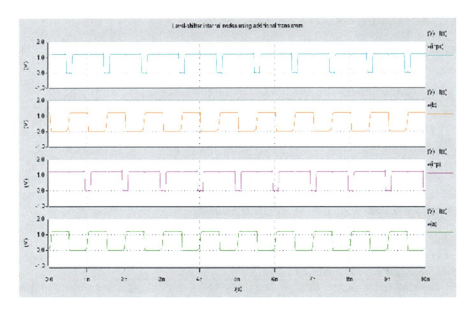

Fig. 3.46 View of signals at interconnection points using a correction system

3.2 Design Principles of Signal Transmission Calibration Systems...

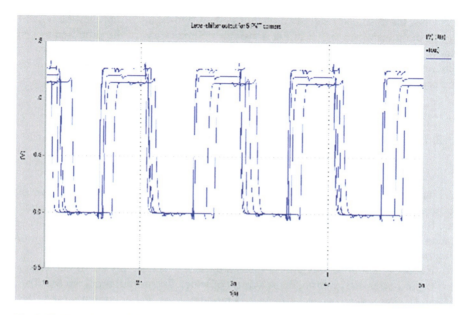

Fig. 3.47 Five view of high-frequency LS output signal under five PVT conditions

Table 3.5 Simulation results using correction system

Parameter	PVT				
	Typical, 1.2 V, 25 °C	Fast, 1.26 V, 125 °C	Fast, 1.26 V, −40 °C	Slow, 1.14 V, 125 °C	Slow, 1.14 V, −40 °C
Duty cycle (%)	49.8	50.2	50.5	49.55	49.7
Jitter (ps)	1.67	1.67	4.16	3.75	2.5

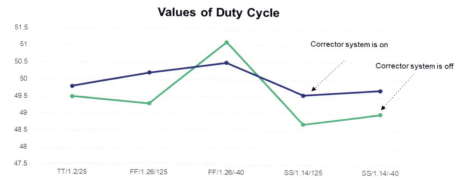

Fig. 3.48 Comparison of duty cycle values

Thus, the designed rise/fall correction system significantly improves the reliability of the transmitted signal in the transmitter. Simulations of the proposed system showed that it calibrates the duty cycle deviations by 39.5% and also reduces the jitter by about twice. However, it increases the area of the high-frequency LS in the transmitter by 12%.

3.2.3 Method of Calibration of Asymmetries of Rise/Fall Times of High-frequency Signals

As mentioned in Sect. 3.1, when the speed of the transmitted signal increases, the calibration of data deviations becomes difficult. The main sources of these deviations are PVT deviations, supply voltage fluctuations, as well as noise. All the mentioned factors cause asymmetries of the transmitted signal rise/fall times, which significantly reduce the reliability of I/O cells (Fig. 3.49).

In order to eliminate deviations, CSSs are used, which supply additional current to the pre-buffer (PB) sub-nodes, as a result of which the rise/fall times are equalized [104]. Deviations can occur during the operation of I/O cell, but currently used CSSs do not calibrate the asymmetries of rise/fall times during data transmission.

The proposed rise/fall asymmetry calibration system allows correction of deviations during the entire operation of the transmitter (Fig. 3.50). The transmitted data is divided into two taps, one of which connects to the calibration system and the other to the PB.

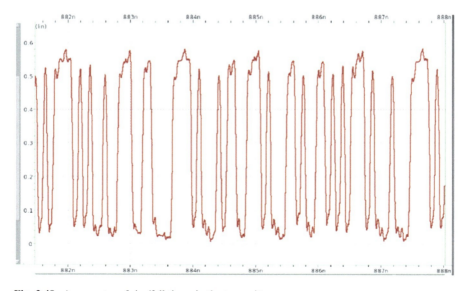

Fig. 3.49 Asymmetry of rise/fall times in the transmitter

3.2 Design Principles of Signal Transmission Calibration Systems...

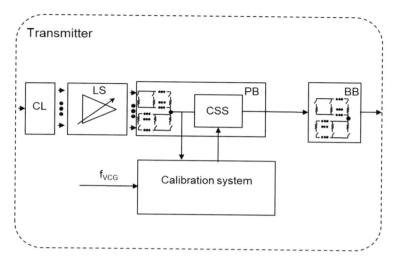

Fig. 3.50 Placement of the calibration system in the transmitter

This structure allows uninterrupted data transfer and calibration during I/O cell operation. The calibration system is clocked by the f_{VCO} signal supplied from the voltage-controlled oscillator (VCO) in the IC. The purpose of the system is to detect the asymmetry of rise/fall times during data transmission, after which to provide additional current to the output node of the PB. It should be noted that deviations can occur both in the case of increasing and decreasing rise/falls. For this reason, the system detects in advance the direction of deviation of rise/falls and amplifies the NMOS or PMOS cascades of the PB output and calibration rise/falls in the given direction. The main components of calibration system (Fig. 3.51) are the input logic cell, integrator, comparators, and logic cell.

The input logic cell separates one period from the transmitted signal, which is then processed to perform the rise/fall time duration calculation. The processing of that signal is carried out by means of an integrator (Fig. 3.52), which slows down the part of one period of the transmitted signal and allows to calculate the duration of rise/fall times.

The slowed signal is supplied to two comparators, the reference voltage of one of which corresponds to 90% of the supply voltage, and the other to 10%. This allows forming signals informing about the start and end of switching of the input signal at the output of the comparators (Fig. 3.53).

For the comparison of high-frequency signals, the clocked comparator (Fig. 3.54) was used in the calibration system [105].

Signals informing about the start and end of switching of the input signal are supplied to a logic cell that counts the durations of rise/fall times and forms the PMOS and NMOS control signals. The latter are connected to the output transistors of PB. They turn off or on the PB's output PMOS or NMOS cascades, resulting in changes in the rise/fall speeds. The logic cell (Fig. 3.55) calculates the speed of the

146 3 Signal Transmission Calibration Systems in Integrated Circuits

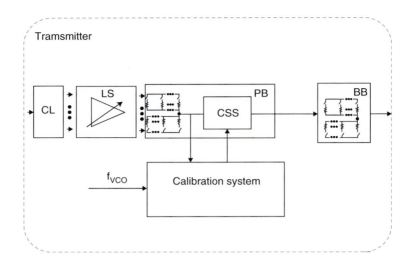

Fig. 3.51 Calibration system structure

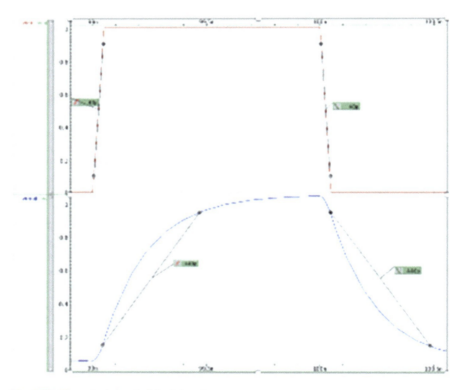

Fig. 3.52 The output signal of the integrator

3.2 Design Principles of Signal Transmission Calibration Systems...

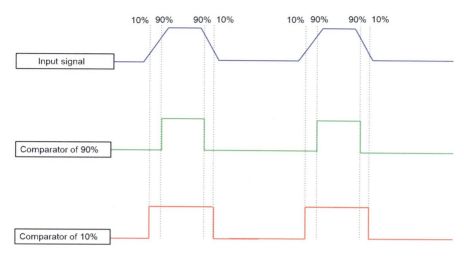

Fig. 3.53 The output signals of the comparators

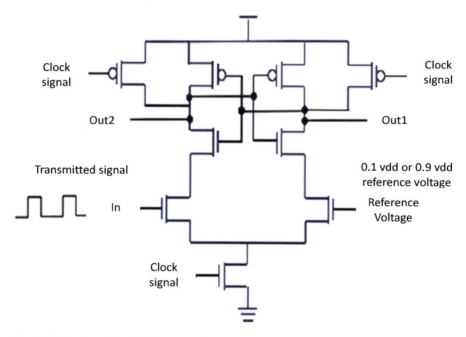

Fig. 3.54 Structure of a clocked comparator

input rise/falls and registers it in registers, after which it detects the direction of deviation using digital comparators. Then the control node supplies the PMOS and NMOS control signals, respectively. This operation is performed until the rise/fall durations of the transmitted signal are equal.

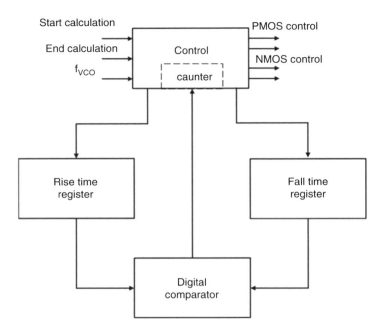

Fig. 3.55 The output signals of comparators

It should be noted that the calculation of the speed of the rise/fall times of the transmitted signal occurs through f_{VCO} signal, increasing the frequency of which will improve the accuracy of the system. The frequency selection of the f_{VCO} signal was made taking into account the integration duration of the signal.

The speed should be sufficient to perform the calculation during the change of rise/fall times.

Thus, with the help of the proposed calibration system, it is possible to correct the asymmetries of rise/fall times during the signal transmission, reducing jitter of the signal.

Evaluation of Effectiveness of Calibration Method of Asymmetries of Rise/Fall Times of High-frequency Signals

A high-frequency transmitter was designed to evaluate the effectiveness of the calibration system for the asymmetry of signal rise/fall times. The simulations were carried out in 27 PVT conditions using a signal corresponding to the DDR4 standard. It should be noted that for the evaluation of the performance of the calibration system, the rise/falls of the input signal of the transmitter were pre-biased in order to simulate the real deviations. Input signals containing significant deviations were applied to the input of the transmitter, which are presented in Table 3.6.

3.2 Design Principles of Signal Transmission Calibration Systems...

Table 3.6 Initial system simulation results

Parameter	PVT		
	Typical, 1.2 V, 25 °C	Fast, 1.26 V, 125 °C	Slow, 1.14 V, −40 °C
Rise/fall asymmetry (ps)	15	24.8	32

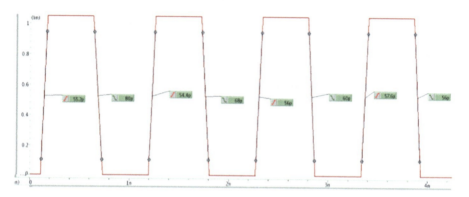

Fig. 3.56 The process of calibration of a signal containing deviations

In the case of slow PVT, the asymmetry of the initial rise/fall times of the input signal of the transmitter was 24.8 ps (Fig. 3.56), which is quite a large bias and can cause significant oscillation and jitter in the back buffer (BB).

It can be noticed that the asymmetry of the rise/fall times of the input signal is corrected during three periods through the calibration system, making only 1.6 ps (Fig. 3.57). It should be noted that the operation of the system does not depend on the direction of deviations, and it does not interrupt the process of transmission of signals.

The calibration accuracy of rise/fall asymmetry can be increased by selecting the sizes of additional PMOS and NMOS transistors in the output node of the PB. It should be noted that the time for the calibration process is different in the case of other PVTs, but it does not exceed five periods. The values of calibrated rise/falls as a result of the system application are presented in Table 3.7.

A comparison with modern rise/fall asymmetry calibration systems was also carried out, which confirmed the effectiveness of the proposed system (Table 3.8).

Thus, the calibration system of designed rise/fall asymmetries significantly reduces the effect of PVT deviations, as a result of which the reliability of the transmitted signal increases. Simulations of the proposed system showed that it calibrates the asymmetries of the signal rise/falls, reaching it to 1.2%, and also reduces jitter. However, it increases the current consumed in the PB node, which is 2.26 mA for the worst case.

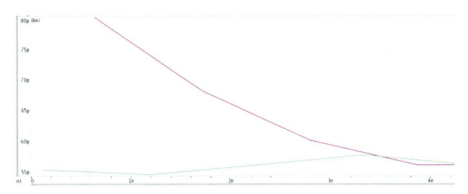

Fig. 3.57 Asymmetry values of input signal rise/falls

Table 3.7 Simulation results without using the system

	PVT		
Parameter	Typical, 1.2 V, 25 °C	Fast, 1.26 V, 125 °C	Slow, 1.14 V, −40 °C
Rise/fall asymmetry (ps)	0.8	1.3	1.6

Table 3.8 Comparison of the proposed calibration system with other systems

Parameter	[9]	[10]	Proposed system
Manufacturing process (μm)	0.13	0.18	0.032
Speed (Mb/s)	630	500	2133
Rise/fall asymmetry (%)	0.9	3.4	1.2

3.2.4 Calibration Method of Signal Distortion Caused by Transmission Line

As mentioned in Sect. 3.1, the transmission line significantly suppresses the transmitted signal, causing inaccuracies and data losses (Fig. 3.58). It should be noted that, even with a coherent line, the transmission line still suppresses the signal, reducing its amplitude and also slowing down the speed of the signal rise/fall times.

The proposed system (Fig. 3.59) [106] reduces the effect of the transmission line. It supplies additional current to the output buffers, as a result of which the signal rise/fall speeds are improved, and it becomes possible to neutralize the losses caused by the transmission line. The designed system is applicable in high-frequency I/O cells and does not interrupt the data transmission process in the transmitter.

In order to carry out signal calibration in the proposed system, a slew rate control amplifier (Fig. 3.60) [107] was designed, which was implemented using an "XOR" logic cell and a frequency divider. It is placed in the node of the transmitter, and during the signal transmission it carries out the amplification of its output signal.

3.2 Design Principles of Signal Transmission Calibration Systems... 151

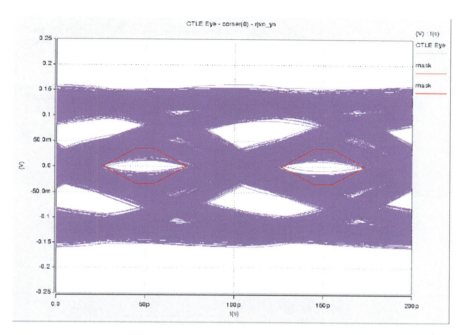

Fig. 3.58 Signal distortions, caused by transmission line

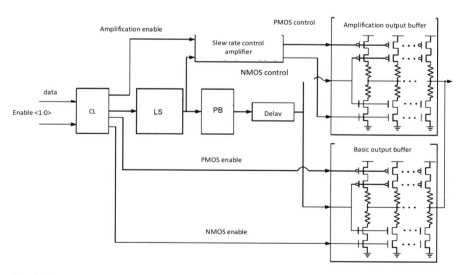

Fig. 3.59 Application of the system at the transmitter

The input signal is applied to one of the inputs of XOR, and its delayed version to the other. As a result, a pulse signal is formed at the output of the "XOR," which gets a logical 1 value during rise/fall times of the input signal (Fig. 3.61) [105]. It should

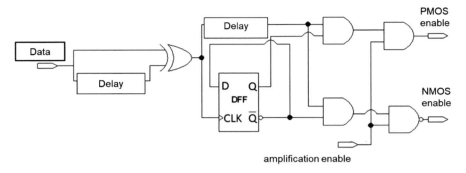

Fig. 3.60 Signal slew rate control system

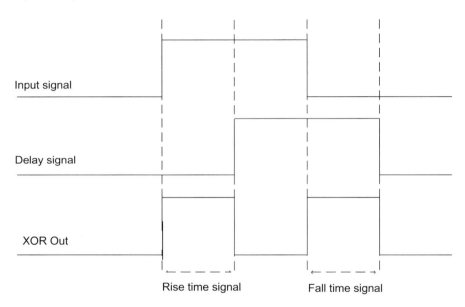

Fig. 3.61. XOR output signal

be noted that the signal delay node was implemented using sequentially connected buffers.

Since the XOR should work at high signal frequencies and have a low delay, the following structure of XOR was chosen (Fig. 3.62) [107].

In order to ensure amplification of PMOS and NMOS nodes in BB, it is necessary to divide the output signal of "XOR" into two taps (Figs. 3.63 and 3.64) [107].

For this purpose, the frequency of the output signal of the XOR is reduced twice by means of a DFF. As a result of this, two signals are formed: PMOS and NMOS amplification, through which additional current is supplied to the BB during

3.2 Design Principles of Signal Transmission Calibration Systems... 153

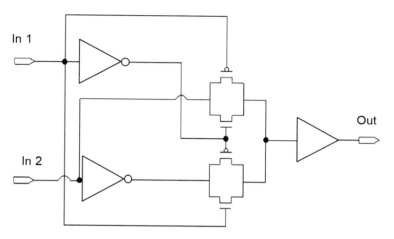

Fig. 3.62 XOR structure

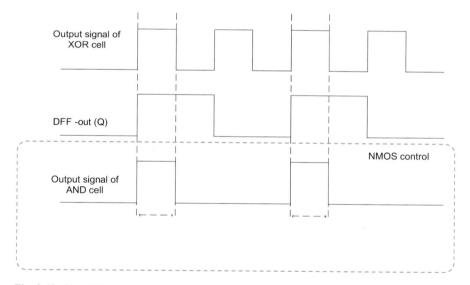

Fig. 3.63 Signal formation of NMOS control

switching of the input signal. It should be noted that the proposed signal supplies additional current from the BB to the transmission line only during signal transitions and is switched off in its static regions to avoid consuming additional current. It is also possible to control the operation of the proposed system and BB using the enable signal <1:0>, and if necessary, turn them off (Table 3.9).

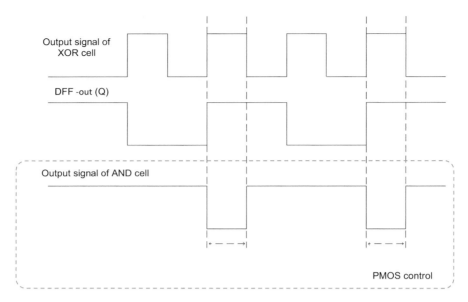

Fig. 3.64 Signal formation of PMOS control

Table 3.9 The working principle of the proposed system

Enable <1:0>	BB operation	The operation of slew rate control system
00	Off	Off
01	Off	On
10	–	–
11	On	On

Evaluation of Effectiveness of Calibration Method of Signal Distortion Caused by Transmission Line

To evaluate the effectiveness of the proposed system, the high-frequency transmitter and slew rate control system were designed. The design was carried out for a 32/28 nm technology, and simulations were also carried out under basic PVT conditions. They were implemented using a 1.2 V supply voltage corresponding to the DDR4 [106] standard and a 1200 MHz data transfer rate. It should be noted that the output resistance of the BB corresponding to the DDR4 standard is equal to 32 Ω, and the values of the two matching resistors at the end of the transmission line are 120 Ω (Fig. 3.65). Two cases were compared to evaluate the efficiency of the system—the signal passes through the transmission line when the system is off and when it is on.

When the system is off, the transmission line significantly suppresses the amplitude of the transmitted signal, as well as the speed of its rise/fall times. In particular,

3.2 Design Principles of Signal Transmission Calibration Systems...

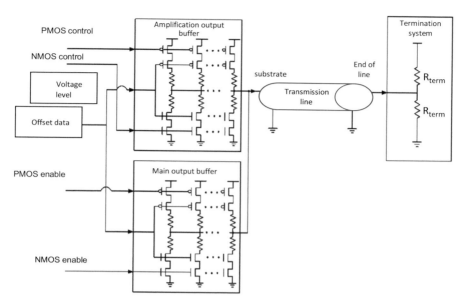

Fig. 3.65 Structure of a system designed for simulation

for the typical case, the high constant voltage level of the signal decreases by 200 mV from the target value (Fig. 3.66).

For other PVTs, the signal amplitude is also suppressed, and for the slow case, the DC signal voltage level is reduced by 260 mV (Fig. 3.67).

Measurements of signal jitter and the speed of rise/fall times were also carried out when the proposed system was turned off (Table 3.10).

Thus, it can be noticed that the structure of BBs currently used significantly limits the further increase in signal transmission speed. As a result of using the proposed system, it was possible to reduce the influence of the transmission line and improve the amplitude of the transmitted signal and the speeds of rise/fall times (Fig. 3.68).

It can be noticed that when the system is used, the speed of the signal rise/fall times increases drastically, and the signal approaches the target value of its constant voltage level. The signal also improves in the case of other PVT conditions (Fig. 3.69).

It should be noted that the system connects additional resistors to the BB during switching, only then supplying additional current from the BB to the transmission line. In the static areas of the signal, this current is absent, as a result of which the consumed current decreases. Jitter measurements were also carried out, which confirmed that the signal jitter is significantly reduced due to the proposed system (Table 3.11).

The use of the system also helps to increase the opening of the eye diagram of the signal (Figs. 3.70 and 3.71). Its horizontal opening increases by 13% and vertical by 10%.

156   3 Signal Transmission Calibration Systems in Integrated Circuits

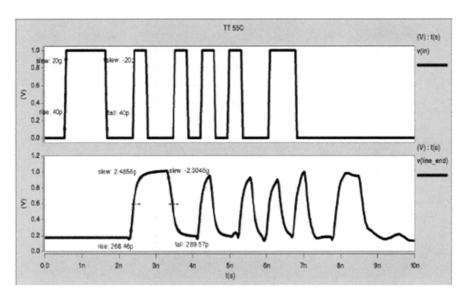

Fig. 3.66 Typical signal appearance without system application

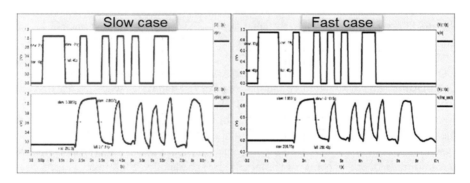

Fig. 3.67 The signal transmitted through the BB in slow and fast cases

Table 3.10 BB simulation results

Parameter	PVT		
	Typical, 1.2 V, 25 °C	Fast, 1.26 V, 125 °C	Slow, 1.14 V, −40 °C
Jitter (ps)	65.5	48.7	79.5
Rise time speed (V/ns)	2.53	3.13	2.02
Fall time speed (V/ns)	2.45	3.03	2.21

3.2 Design Principles of Signal Transmission Calibration Systems...

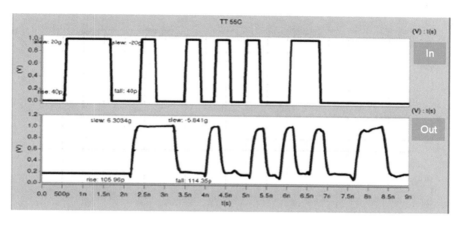

Fig. 3.68 Typical signal view using the proposed system

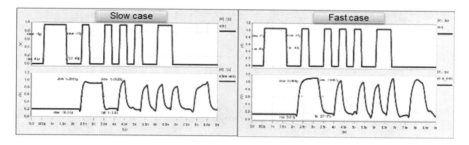

Fig. 3.69 The transmitted signal using the proposed system in slow and fast cases

Table 3.11 Simulation results using the proposed system

Parameter	PVT		
	Typical, 1.2 V, 25 °C	Fast, 1.26 V, 125 °C	Slow, 1.14 V, −40 °C
Jitter (ps)	51.3	35.7	63.1
Rise time speed (V/ns)	6.62	6.81	5.41
Fall time speed (V/ns)	6.02	6.48	5.15

Thus, through the designed calibration system, it is possible to reduce the effect of the transmission line, as a result of which the speed of the signal rise/fall times increases by 50%, the horizontal and vertical openings of the signal eye at the end of the transmission line increase by 13% and 10%, respectively, as well as signal jitter is reduced by 20.7%. However, the proposed system increases the area of BB by 13%.

158   3 Signal Transmission Calibration Systems in Integrated Circuits

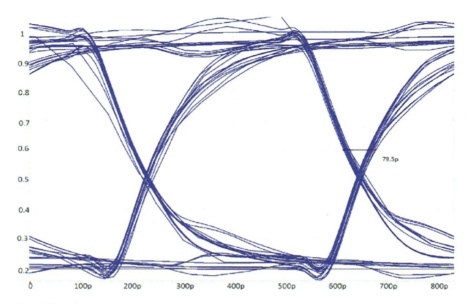

Fig. 3.70 Using BB, an eye diagram of a signal at the end of the transmission line

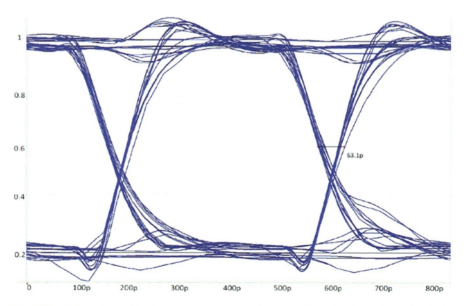

Fig. 3.71 Using the calibration system, an eye diagram of a signal at the end of the transmission line

Conclusion

1. Principles of development of signal transmission calibration means in integrated circuits have been proposed, which allow to significantly improve their main technical characteristics and parameters: speed, reliability of data transmission and reading, consumed power, etc.
2. A self-calibration method for detection of deviations in the duty cycle of the signal has been developed, in which, due to reading the data through the applied digital nodes, it significantly improves the reliability of the data and the calibration duty cycle with an accuracy of ±0.5% at the expense of the increase of consumed power of the receiver by only 3.95 mW.
3. A method of increasing the reliability of high-frequency data in the sub-nodes of the transmitter was proposed, which, due to the supply of additional current, calibrates the deviations of the duty cycle in the voltage converter by 39.5% and also reduces the jitter twice at the expense of increasing the area of the high-frequency voltage converter in the transmitter by 12%.
4. A calibration method for the asymmetry of high-frequency signal rise/falls has been developed, which significantly reduces the effect of process-voltage-temperature deviations, as a result of which the asymmetry of the rise/falls is calibrated, reaching 1.2%; the jitter is also reduced at the expense of the pre-buffer, increasing the consumed current by 2.26 mA.
5. A transmission line-induced signal distortion calibration method was proposed, which reduces the effect of the transmission line, as a result of which the speed of the transmitted signal rise/fall times increases by 50%, the horizontal and vertical openings of the signal eye increase by 13% and 10%, respectively, as well as the signal jitter is reduced by 20.7% due to the 13% increase in output buffer area.

References

1. G. Madrigal-Boza, M. Oviedo-Hernández, A. Carmona-Cruz, et al., An IC mixed-signal framework for design, optimization, and verification of high-speed links. IEEE 11th Latin American symposium on circuits & systems (LASCAS-2020) (2020), pp. 1–4
2. http://irufa.blogspot.com/2011/11/computer-hardware-basics-computer.html
3. M. Lee, W. Dally, P. Chiang, Low-power area-efficient high-speed I/O circuit techniques. IEEE J. Solid State Circuits **35**, 1591–1599 (2000)
4. A. Agrawal, *Design of High Speed I/O Interfaces for High Performance Microprocessors* (Harvard University, Cambridge 2010), 119 p
5. Jedec DDR standards & specification document. June, 2017
6. N. Rodriguez, F. Gamiz, S. Cristoloveanu, A-RAM memory cell: Concept and operation. IEEE Electron. Device Lett. **31**, 972–974 (2010)
7. Jedec LPDDR standards & specification document. December, 2018
8. J. Axelson, *USB Complete: The Developer's Guide* (Lakeview Research LLC, Madison, 2015)
9. H. Jun, J. Cho, K. Lee, Y. Son, et al., Hbm (high bandwidth memory) dram technology and architecture. IEEE international memory workshop (IMW-2017) (2017), pp. 1–4

10. B. Dingle, J. Eubanks, K. Janasak, 3D RAM modeling and simulation in a model based systems engineering environment. IEEE annual reliability and maintainability symposium (RAMS-2020) (2020), pp. 1–6
11. S. Eidson, B. Gaines, P. Wolf, 30.2: HDMI: High-definition multimedia Interface, in *SID Symposium Digest of Technical Papers*, (Blackwell Publishing Ltd, Oxford, UK, 2003), pp. 1024–1027
12. A. Sedzin, J. Aguilar, A. Marechal, R. O'Connor, et al., High-speed inter-IC interfacing for mobile multimedia applications, in *Digest of Technical Papers International Conference on Consumer Electronics*, (IEEE, Piscataway, 2007), pp. 1–2
13. A.S. Trdatyan, *Development of Self-Configurable Input/Output Units for Integrated Circuits*. PhD dissertation. Yerevan, (2020), 162 pages
14. S. Chun, M. Swaminathan, L. Smith, et al., Modeling of simultaneous switching noise in high speed systems. IEEE Trans. Adv. Packag. **24**, 132–142 (2001)
15. A. Fish, V. Milrud, O. Yadid-Pecht, High-speed and high-precision current winner-take-all circuit. IEEE Trans. Circuits Syst. II Express Briefs **52**, 131–135 (2005)
16. J. Ardenkjaer-Larsen, B. Fridlund, A. Gram, G. Hansson, et al., Increase in signal-to-noise ratio of > 10,000 times in liquid-state NMR. Proc. Natl Acad. Sci. **100**, 10158–10163 (2003)
17. N. Rao, G. Knight, S. Mohan, et al., Studies on failure of transmission line towers in testing. Eng. Struct. **35**, 55–70 (2012)
18. A. Djordjevic, A. Zajic, G. Tosic, et al., A note on the modeling of transmission-line losses. IEEE Trans. Microwave Theory Techz **51**, 483–486 (2003)
19. https://www.electronicdesign.com/technologies/communications/article/21796367/back-to-basics-impedance-matching-part-1
20. E. Turan, S. Demir, An all 50ohm divider/combiner structure. IEEE MTT-S international microwave symposium digest (Cat. No. 02CH37278) (2002), pp. 105–108
21. A. Mangan, S. Voinigescu, M. Yang, et al., De-embedding transmission line measurements for accurate modeling of IC designs. IEEE Trans. Electron. Devices **53**, 235–241 (2006)
22. V. Melikyan, A. Balabanyan, A. Hayrapetyan, et al., Receiver/transmitter input/output termination resistance calibration method. IEEE XXXIII international scientific conference electronics and nanotechnology (ELNANO-2013) (2013), pp. 126–130
23. H. Johnson, M. Graham, *High-Speed Signal Propagation: Advanced Black Magic* (Prentice Hall Professional, Upper Saddle River, 2003), 808p.
24. S. Lehmann, F. Gerfers, Channel analysis for a 6.4 Gb/s DDR5 data buffer receiver front-end. 15th IEEE international new circuits and systems conference (NEWCAS-2017) (2017), pp. 109–112
25. K. Khachikyan, L. Msryan, A. Balabanyan, Research of PVT variation influence on PLL system and methodology of control voltage stabilization. IEEE 37th international conference on electronics and nanotechnology (ELNANO-2017) (2017), pp. 190–193
26. V. Melikyan, K. Khachikyan, H. Gumroyan, et al., Crystal area reduction method for impedance matching systems in high-speed data links. Proc. Univ. Electron. **24**(5), 503–510 (2019)
27. V. Melikyan, A. Hayrapetyan, B. Baghramyan, et al., Transmitter output impedance calibration method. IEEE east-west design & test symposium (EWDTS-2018) (2018), pp. 1–8
28. W. Bae, Supply-scalable high-speed I/O interfaces. Electronics **9**, 1315 (2020)
29. L. Fassio, F. Settino, L. Lin, R. De Rose, et al., A robust, high-speed and energy-efficient ultralow-voltage level shifter. IEEE Trans. Circuits Syst. II Express Briefs **68**, 1393–1397 (2020)
30. B. Mahendranath, A. Srinivasulu, Output buffer for+ 3.3 V applications in a 180 nm+ 1.8 V CMOS technology. Radioelectron. Commun. Syst. **60**, 512–518 (2017)
31. H. Yu, T. Michalka, M. Larbi, M. Swaminathan, Behavioral modeling of tunable I/O drivers with preemphasis including power supply noise. IEEE Trans. Very Large-Scale Integr Syst. **28**, 233–242 (2019)

References

32. M. Ker, T. Wang, F. Hu, Design on mixed-voltage I/O buffers with slew-rate control in low-voltage CMOS process. IEEE 15th international conference on electronics, circuits and systems (2008), pp. 1047–1050
33. https://www.sciencedirect.com/topics/engineering/characteristic-impedance
34. J.F. Wakerly, Transmission lines, reflections, and termination, in *Digital Design Principles and Practices*, 4th edn. (Pearson Education, Inc., Upper Saddle River, 2006), 112 p
35. A.J. Deutsch, W. Res, G.V. Kopcsay, et al., When are transmission-line effects important for on-chip interconnections. Microwave Theory Tech. IEEE Trans. RFIC Virtual J. **45**(10), 1836–1846 (1997)
36. V. Melikyan, A. Balabanyan, A. Hayrapetyan, A. Durgaryan, NMOS/PMOS resistance calibration method using reference frequency. IEEE ninth international conference on computer science and information technologies revised selected papers (2013), pp. 1–6
37. A. Balabanyan, A. Durgaryan, Fully integrated PVT detection and impedance self-calibration system design. IEEE XXV international scientific conference electronics (ET-2016) (2016). pp. 1–4
38. V. Melikyan, A. Sahakyan, A. Hayrapetyan, et al., Serializer/deserializer output data signal duty cycle correction method. Proceedings of 57th ETRAN conference (2013), pp. 1–4
39. J. Chung, A.A. Iliadis, Design and optimization of a CMOS IC novel RF tracking sensor. Int. J. Circuit Theory Appl. **49**, 801–819 (2021)
40. O.H. Petrosyan, A.A. Martirosyan, A.S. Trdatyan, et al., Equalization method of resistors. Manual Eng. Acad. Armenia Yerevan **15**(3), 475–479 (2018) (in Armenian)
41. B. Sporrer, L. Wu, L. Bettini, et al., A fully integrated dual-channel on-coil CMOS receiver for array coils in 1.5–10.5 T MRI. IEEE Trans. Biomed. Circuits Syst. **11**, 1245–1255 (2017)
42. Y. Lai, Y. Liao, J. Jou, et al. Design of high-speed optical receiver module for 160Gb/s NRZ and 200Gb/s PAM4 transmissions. IEEE international symposium on circuits and systems (ISCAS-2019) (2019), pp. 1–4
43. B. Fahs, J. Chowdhury, M. Hella, A 12-m 2.5-Gb/s lighting compatible integrated receiver for OOK visible light communication links. J. Lightwave Technol. **34**, 3768–3775 (2016)
44. G. Li, Y. Yin, Y. Zhang, High-precision mixed modulation DAC for an 8-bit AMOLED driver IC. J. Disp. Technol. **11**, 423–429 (2015)
45. J. Jun, J. Kang, S. Kim, A 16-bit incremental ADC with swapping DAC for low power sensor applications. IEEE international symposium on circuits and systems (ISCAS-2019) (2019), pp. 1–4
46. V. Melikyan, K. Khachikyan, A. Matevosyan, A. Petrosyan, et al., High quality factor 5.0 Gbps CTLE circuit for SERDES serial links. IEEE East-west design & test symposium (EWDTS-2018) (2018), pp. 1–5
47. V. Melikyan, A. Balabanyan, A. Durgaryan, et al., PVT variation detection and compensation methods for high-speed systems. IFIP/IEEE 21st international conference on very large-scale integration (VLSI-SoC-2013) (2013), pp. 322–327
48. V. Melikyan, K. Khachikyan, A. Trdatyan, A. Petrosyan, et al., High quality factor 5.0 Gbps CTLE circuit for SERDES serial links. IEEE east-west design & test symposium (EWDTS), Kazan, Russia, 14 September 2018 (2018), pp. 641–644
49. V.K. Sharma, S. Deb, *Analysis and Estimation of Jitter Sub-components*. Dissertation (IIIT, Delhi, 2014), 96 p
50. N. Kirianaki, Y. Yurish, *Data Acquisition and Signal Processing for Smart Sensors* (Wiley, Newark, 2002), 274 p
51. J. Fan, X. Ye, J. Kim, B. Archambeault, et al., Signal integrity design for high-speed digital circuits: Progress and directions. IEEE Trans. Electromagn. Compat. **52**, 392–400 (2010)
52. S. Ooi, L. Kong, H. Goay, et al., Crosstalk modeling in high-speed transmission lines by multilayer perceptron neural networks. Neural Comput. & Applic. **32**, 7311–7320 (2020)
53. K. Khachikyan, A. Balabanyan, A. Petrosyan, PLL control voltage stabilization method for high-speed systems. IEEE XXV international scientific conference electronics (ET-2016) (2016), pp. 1–4

54. M. Shinagawa, Y. Akazawa, T. Wakimoto, Jitter analysis of high-speed sampling systems. IEEE J. Solid State Circuits **25**, 220–224 (1990)
55. Shimanouchi M., An approach to consistent jitter modeling for various jitter aspects and measurement methods. IEEE proceedings international test conference (Cat. No. 01CH37260) (2001), pp. 848–857
56. S. Boscolo, J. Fatome, C. Finot, Impact of amplitude jitter and signal-to-noise ratio on the nonlinear spectral compression in optical fibres. Opt. Commun. **389**, 197–202 (2017)
57. J. Yamaguchi, M. Soma, M. Ishida, et al., Extraction of peak-to-peak and RMS sinusoidal jitter using an analytic signal method. Proceedings 18th IEEE VLSI test symposium (2000), pp. 395–402
58. A. Rysin, P. Livshits, S. Sofer, et al. Inter-symbol interference (ISI) in on-die transmission lines. IEEE international conference on microwaves, communications, antennas and electronics systems (2009), pp. 1–5
59. J.F. Buckwalter, *Deterministic Jitter in Broadband Communication*. Dissertation (California Institute of Technology, Pasadena, 2006), 186 p
60. J. Schoentgen, R. De Guchteneere, Predictable and random components of jitter. Speech Comm. **21**, 255–272 (1997)
61. K. Bhattacharyya, S. Xu, S. Bhattacharya, Impact on inter symbol interference (ISI) noise due to simulation error. IEEE 14th topical meeting on electrical performance of electronic packaging (2005), pp. 221–224
62. T. Wong, *Theory of Digital Communications* (University of Florida, Gainesville, 2006), 81 p
63. V. Melikyan, K. Khachikyan, A. Trdatyan, A. Martirosyan, et al., Process variation detection and self-calibration method for high-speed serial links. IEEE east-west design & test symposium (EWDTS-2018) (2018), pp. 1–4
64. B. Razavi, Challenges in the design high-speed clock and data recovery circuits. IEEE Commun. Mag. **40**, 94–101 (2002)
65. N. Baptistat, K. Abouda, G. Duchamp, T. Dubois, et al., Effects of process-voltage-temperature (PVT) variations on low-side MOSFET circuit conducted emission. IEEE 12th international workshop on the electromagnetic compatibility of integrated circuits (EMC Compo-2019) (2019), pp. 213–215
66. X. Guan, T. Yang, F. Tang, A 5-GHz phase compensation spread spectrum clock generator for high speed SerDes application. IEEE 5th international conference on integrated circuits and microsystems (ICICM-2020) (2020), pp. 293–297
67. K. Venkatachala, S. Leuenberger, A. ElShater, et al., Passive compensation for improved settling and large signal stabilization of ring amplifiers. IEEE international symposium on circuits and systems (ISCAS-2018) (2018), pp. 1–5
68. S. Gajare, K. Pradhan, V. Terzija, A method for accurate parameter estimation of series compensated transmission lines using synchronized data. IEEE Trans. Power Syst. **32**, 4843–4850 (2017)
69. J. Lee, Y. Tsai, W. Lin, et al., A slew rate variation compensated 2 times VDD I/O buffer using deterministic P/N-PVT variation detection method. IEEE Trans. Circuits Syst. II Express Briefs **66**, 116–120 (2018)
70. X. Song, J. Fang, B. Han, et al., Adaptive compensation method for high-speed surface PMSM sensorless drives of EMF-based position estimation error. IEEE Trans. Power Electron. **31**, 1438–1449 (2015)
71. F. Jhong, P. Pan, H. Cheng, et al., Improving high-speed signal transmission loss by low conductor surface roughness. IEEE 17th electronics packaging and technology conference (EPTC-2015) (2015), pp. 1–4
72. F. Jhong, P. Pan, H. Cheng, et al., Measurement analysis and improvement technique of signal integrity for high-speed connectors. Asia-Pacific symposium on electromagnetic compatibility (2012), pp. 609–612
73. J. Ardizzoni, High-speed time-domain measurements – Practical tips for improvement. Analog Dialogue **41**, 13–18 (2007)

References

74. A. Sidorov, N. Goryunov, S. Golubkov, Improvement of automatic control system for high-speed current collectors. J. Phys. Conf. Ser. **944**, 012108 (2018) IOP Publishing
75. A. Malkov, D. Vasiounin, O. Semenov, A review of PVT compensation circuits for advanced CMOS technologies. Circuits Syst. **2**, 162–169 (2011)
76. L. Tan, K. Chan, A fully integrated point-of-load digital system supply with PVT compensation. IEEE Trans. Very Large-Scale Integr. Syst. **24**, 1421–1429 (2015)
77. M. Marcu, S. Durbha, S. Gupta, Duty-cycle distortion and specifications for jitter test-signal generation. IEEE international symposium on electromagnetic compatibility (2008), pp. 1–4
78. M. Bushnell, V. Agrawal, *Essentials of Electronic Testing for Digital, Memory and Mixed-Signal VLSI Circuits* (Springer, New York, 2004), 574 p
79. R. Mehta, S. Seth, S. Shashidharan, B. Chattopadhyay, et al., A programmable, multi-GHz, wide-range duty cycle correction circuit in 45nm CMOS process. IEEE proceedings of the ESSCIRC (2012), pp. 257–260
80. J. Melo, N. Paulino, J. Goes, Continuous-time delta-sigma modulators based on passive RC integrators. IEEE Trans. Circuits Syst. I Regul. Pap. **65**, 3662–3674 (2018)
81. R. Hosseini, M. Saberi, R. Lotfi, A high-speed and power-efficient voltage level shifter for dual-supply applications. IEEE Trans. Very Large-Scale Integr. Syst. **25**, 1154–1158 (2016)
82. Razavi B., *Fundamentals of Microelectronics* (Wiley, Hoboken, 2021), 928 p
83. K. Sharma, N. Tripathi, R. Nagpal, et al., A comparative analysis of jitter estimation techniques. IEEE international conference on electronics, communication and computational engineering (ICECCE-2014) (2014), pp. 125–130
84. N. Tripathi, K. Sharma, H. Advani, et al., An analysis of power supply induced jitter for a voltage mode driver in high speed serial links. IEEE 20th workshop on signal and power integrity (SPI-2016) (2016), pp. 1–4
85. Y. Shim, D. Oh, T. Hoang, et al., A jitter equalization technique for minimizing supply noise induced jitter in high speed serial links. IEEE international symposium on electromagnetic compatibility (EMC-2014) (2014), pp. 827–832
86. S. Valadimas, Y. Tsiatouhas, A. Arapoyanni, Timing error tolerance in nanometer ICs. IEEE 16th international on-line testing symposium (2010), pp. 283–288
87. S. Lee, A. Saad, L. Lee, et al., On-chip slew-rate control for low-voltage differential signalling (LVDS) driver. IEEE international symposium on intelligent signal processing and communication systems (ISPACS-2014) (2014), pp. 99–101
88. K. Szczerba, T. Lengyel, M. Karlsson, et al., 94-Gb/s 4-PAM using an 850-nm VCSEL, pre-emphasis, and receiver equalization. IEEE Photon. Technol. Lett. **28**, 2519–2521 (2016)
89. Z. Zhou, T. Odedeyi, B. Kelly, J. O'Carroll, et al., Impact of analog and digital pre-emphasis on the signal-to-noise ratio of bandwidth-limited optical transceivers. IEEE Photonics J. **12**, 1–2 (2020)
90. Y. Lu, K. Jung, Y. Hidaka, et al., Design and analysis of energy-efficient reconfigurable pre-emphasis voltage-mode transmitters. IEEE J. Solid State Circuits **48**, 1898–1909 (2013)
91. Yuminaka Y., Takahashi Y., Time-domain pre-emphasis techniques for equalization of multiple-valued data. IEEE 38th international symposium on multiple valued logic (ISMVL-2008) (2008), pp. 20–25
92. K. Khachikyan, A. Balabanyan, H. Gumroyan, Precise duty cycle variation detection and self-calibration system for high-speed data links. IEEE computer society annual symposium on VLSI (ISVLSI-2018) (2018), pp. 191–196
93. H. Huang, J. Heilmeyer, M. Grözing, M. Berroth, et al., An 8-bit 100-GS/s distributed DAC in 28-nm CMOS for optical communications. IEEE Trans. Microwave Theory Tech. **63**, 1211–1218 (2015)
94. J. Shen, A. Shikata, D. Fernando, et al., A 16-bit 16-MS/s SAR ADC with on-chip calibration in 55-nm CMOS. IEEE J. Solid State Circuits **53**, 1149–1160 (2018)
95. Y. Delican, T. Yildirim, High performance 8-bit mux-based multiplier design using MOS current mode logic. IEEE 7th international conference on electrical and electronics engineering (ELECO-2011) (2011), pp. 89–93

96. L. Melo, J. Goes, N. Paulino, A 0.7 V 256 μW ΔΣ modulator with passive RC integrators achieving 76 dB DR in 2 MHz BW. IEEE symposium on VLSI circuits (VLSI circuits) (2015), pp. 290–291
97. V.Sh. Melikyan, M. Martirosyan, A. Melikyan, G. Piliposyan, 14nm educational design kit: Capabilities, deployment and future. Proceedings of the 7th small systems simulation symposium 2018, Niš, Serbia, 12–14 February 2018 (2018), pp. 37–41
98. I. Raja, G. Banerjee, A. Zeidan, et al., A 0.1–3.5-GHz duty-cycle measurement and correction technique in 130-nm CMOS. IEEE Trans. Very Large-Scale Integr. Syst. **24**, 1975–1983 (2015)
99. P. Chen, W. Chen, J. Lai, A low power wide range duty cycle corrector based on pulse shrinking/stretching mechanism. Proc. IEEE Asian solid-state circuits conference (2007), pp. 460–463
100. C. Jang, J. Bae, J. Park, CMOS digital duty cycle correction circuit for multi-phase clock. Electron. Lett. **39**, 1383–1384 (2003)
101. D. Pan, H.W. Li, B.M. Wilamowski, A low voltage to high voltage level shifter circuit for MEMS application. IEEE proceedings of the 15th biennial university/government/industry microelectronics symposium (2003), pp. 128–131
102. V. Melikyan, K. Khachikjan, L. Msryan, et al., High speed, low-jitter level shifter for high speed ICs. IEEE 37th international conference on electronics and nanotechnology (ELNANO-2017) (2017), pp. 175–177
103. R.J. Baker, C. Boyce, *Circuit Design, Layout, and Simulation*, IEEE press series on microelectronic systems (IEEE-Press, Piscataway, 2005), pp. 350–453
104. Y. Ho, H.K. Chen, C. Su, Energy-effective sub-threshold interconnect design using high-boosting predrivers. IEEE J. Emerging Sel. Top. Circuits Syst. **2**, 307–313 (2012)
105. M. Abbas, Y. Furukawa, S. Komatsu, et al., Clocked comparator for high-speed applications in 65nm technology. IEEE Asian solid-state circuits conference (2010), pp. 1–4
106. JEDEC solid state technology association. Low Power Double Data 4 (LPDDR4). November, 2015
107. V. Melikyan, K. Khachikyan, A. Trdatyan, A. Durgaryan, Design of edge boosting digital control circuit for high-speed ICs. IEEE 36th international conference on electronics and nanotechnology (ELNANO-2016) (2016), pp. 315–318

Chapter 4
Methods to Improve Linearity of Signal's Analog-to-Digital Conversion with Self-Calibration

4.1 General Issues to Improve Linearity of Signal's Analog-to-Digital Conversion with Self-Calibration

4.1.1 Importance of Means to Improve the Linearity of Signal's Analog-to-Digital Conversion with Self-Calibration

Nowadays, it is impossible to imagine the operation of modern electronic devices without digital-to-analog (DAC) and analog-to-digital (ADC) converters [1, 2]. DACs and ADCs are fundamental working blocks in many complex integrated circuits (ICs), and the accuracy of their operation directly depends on the whole system [2–4]. The working drawbacks of these blocks can lead to system errors. Therefore, the first task is to design IC as per technical parameters. But in many cases, even a precisely designed circuit does not provide accurate operation after manufacturing, the reason of which is precisely the manufacturing inaccuracies and not the errors, made at the design stage [5–8]. To present the problem in more detail, the simplest DAC, the Kelvin divider or r-string DAC, is considered [9–12]. The N-bit version of this DAC contains 2^N series equal resistors and 2^N switches, typically of complementary metal-oxide-semiconductor (CMOS) construction, connected between the resistors and the output (Fig. 4.1) [9]. Depending on the digital code, the corresponding voltage is transferred through switches at the output, forming an analog signal. One of the advantages of this DAC is the monotonic transfer function, which is caused by the resistors connected in series, and even if one of the resistors is short-connected, the voltage is distributed among the remaining resistors, and the error is not significant. A problem can arise when the values of resistors differ from each other, leading to nonlinear operation of the circuit.

Usually, at the design stage, the values of series resistors are chosen equal, but due to random deviations and inaccuracies as a result of manufacturing process, they

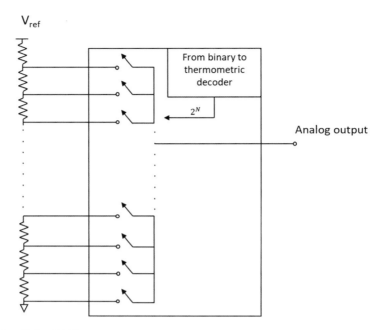

Fig. 4.1 r-String DAC

deviate from the nominal, as a result of which the nonlinearity of the circuit increases [9, 13–16].

The main parameters characterizing the linearity of ADCs and DACs are differential (DNL) and integral (INL) nonlinearities [9]. In an ideal case, when nonlinearities are absent in the operation of DAC, the increment of the digital code should correspond to a voltage increase in the analog output equal to one least significant bit (LSB), and in the case of an ADC, a change in the input voltage of 1 LSB should correspond to a switching from one value of the digital code to the next one. The DNL error is defined as the maximum deviation of the switching from the ideal 1 LSB size in the transfer function. The ideal DAC transfer function without linearity error (Fig. 4.2) has a monotonic view, and the view of the nonlinear DAC transfer function is non-monotonic (Fig. 4.3) [9].

The numerical code is represented in the decimal system. The linear transfer function of the ADC will have an ideal stepwise view, and the transfer function with linearity error will have a non-monotonic view (Fig. 4.4).

A DNL error can lead to missing code and voltage [9]. When DNL is less than one 1 LSB for at least one of the switches, then DNL is considered non-monotonic and the transfer function contains one local maximum or minimum. When DNL is greater than 1 LSB, the problem of monotonicity does not arise, but such a transmission characteristic is also not so desirable (Fig. 4.5) [9].

In many DAC applications, it is unacceptable to have a non-monotonic transfer function, as this can lead to a complete failure of the circuit, especially in cases where

4.1 General Issues to Improve Linearity of Signal's Analog-to-Digital... 167

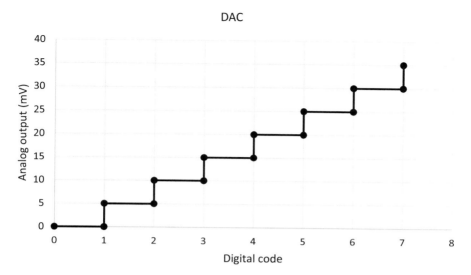

Fig. 4.2 Ideal transfer characteristic of a DAC

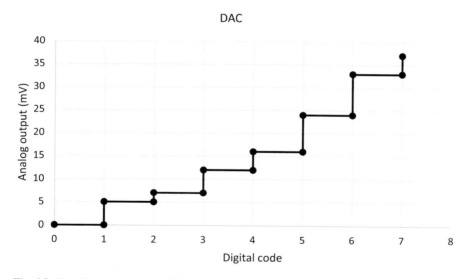

Fig. 4.3 Transfer characteristic of DAC with linearity error

the DAC is part of a negative feedback system, and its non-monotonicity can make the feedback positive. The DNL of the DAC is defined as the deviation of the switching from the ideal 1 LSB size in the transfer function (4.1) [9, 17, 18].

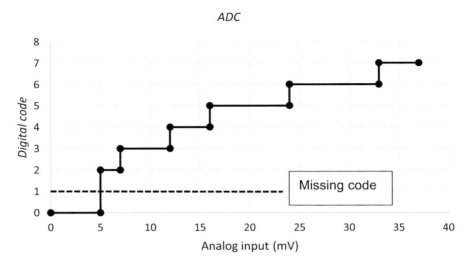

Fig. 4.4 Transfer characteristic of ADC with linearity error

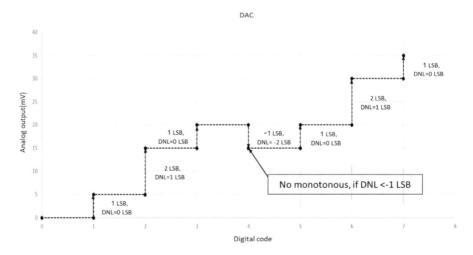

Fig. 4.5 DAC differential nonlinearity details

$$\text{DNL} = \frac{V_{i+1} - V_i}{\text{LSB}} - 1 \qquad (4.1)$$

The INL of a DAC for a given code is the difference between the available voltage and the ideal transfer function voltage (4.2) [9, 17, 18].

4.1 General Issues to Improve Linearity of Signal's Analog-to-Digital...

$$\text{INL} = \frac{V_i - V_0}{\text{LSB}} - i \quad (4.2)$$

An INL error can also lead to code loss. DNL and INL are mainly caused by technological deviations and inaccuracies in the manufacturing process. Usually, the deviation of discrete electrical elements (resistors, capacitors, etc.) from the nominal value after the manufacturing process is not very large and does not exceed 1%, but the situation is different in ICs. In ICs, after manufacturing, the deviation of the parameters of separate cells can reach up to ±25% from the nominal [19, 20]. DACs and ADCs, manufactured with such deviations, have significant nonlinearity errors and often code loss due to a large DNL or INL error.

Using the example of the r-string DAC, the effect of deviations in the physical parameters of electrical components on the system performance caused by manufacturing is considered. In the r-string DAC, such deviations are carried mainly by resistors and switches, which have a CMOS structure. Their parasitic parameters, such as parasitic capacitance and resistance, also deviate from the nominal values during manufacturing process, which in turn affects the delay of switch, leading to inaccuracies in the operation of a circuit.

Thus, even precisely designed ICs, which fully meet specification requirements at design stage, inevitably suffer deterioration of operating parameters as a result of manufacturing, the cause of which is the manufacturing itself. In particular, nonlinearity errors in DACs and ADCs increase, which can lead to system malfunction and as a result reduce the percentage of the output of operating circuits. However, after the end of manufacturing process, it is not possible to correct the operating parameters of the latter, so the development of self-calibration deviation correction systems in the IC is an important issue.

4.1.2 Need for Means to Improve the Linearity of Signal's Analog-to-Digital Conversion with Self-Calibration

The problems discussed and proposed in Sect. 4.1.1, as already mentioned, cannot be completely solved at the design stage. For this reason, it is necessary to develop and introduce such means that will allow correcting errors and deviations in the manufacturing process by self-calibration after manufacturing, to have a precisely operating circuit. The problem is challenging, because the semiconductor industry is one of the fastest growing directions of the economy. The growth of problems in the sector also creates the need to have compatible solutions. Thus, ICs with embedded self-calibration systems are in high demand because, in addition to solving the problem of improving IC operating parameters, they simultaneously increase the number of operating circuits, reducing production costs. However, developing such systems is not an easy task.

- It is very difficult, if not impossible, to have complex testing and measurement systems in an IC, considering area/power consumption ratio.
- Such embedded systems, in turn, have inaccuracies and may cause additional errors. Therefore, the proposed solutions should be as error-free as possible, or the errors should be less significant.
- Testing, measurement, and subsequent self-calibration should be performed in the order of minimum steps, because the more complex the system, the longer the self-calibration time, and the design of such a system requires more resources.

Thus, the development of self-calibration DACs and ADCs means of increasing the linearity is a very necessary and, at the same time, difficult task.

4.1.3 Causes of Nonlinearity Occurrence in Signal's Analog-to-Digital Conversion and the Importance and Necessity of its Reduction in Flash ADCs

Another well-known circuit is discussed, where again due to manufacturing deviations and inaccuracies, deterioration of the quality of the circuit operation up to complete failure can occur. Flash ADC is the fastest and simplest ADC in structure [21–24]. This ADC allows to digitize 2n number of voltage values through 2n-1 number of comparators. At the output, the digital code is generated during one stage. The speed of the flash ADC depends only on comparators and to some extent on a decoder (Fig. 4.6) [21].

The reference voltages, applied to comparators, are obtained by a series resistors. A chain of resistors is connected between positive and negative reference voltages (or between the positive voltage and zero). A 2n-1-bit output code (thermal code) is formed at the output of comparators, which is converted into an N-bit digital output code by means of a decoder with 2n inputs. Out of the 2n inputs of the decoder, only 2n-1 have comparators [21].

The number of comparators used in a flash ADC is exponentially dependent on the number of bits in the output digital code. Therefore, the use of flash ADCs in systems requiring high bit is not very appropriate. In practice, the bit limit of flash ADCs does not exceed 8 bits [25–27]. The main advantage of flash ADC is speed. One of its disadvantages is the large area it occupies [28].

In this circuit, as in the case of the r-string DAC, the manufacturing inaccuracies mainly affect the deviation from the nominal value of resistors, as well as the linearity of comparators, which leads to nonlinearity of flash ADC.

If the deviation of resistors represents a deviation from the nominal value, then the linearity in case of a comparator is due to several factors. To present the problem more clearly, the simplest comparator (Fig. 4.7) [29–33] is considered.

The main function of a comparator is to compare two reference voltages: when one voltage exceeds the other, the output of the comparator switches. Comparators

4.1 General Issues to Improve Linearity of Signal's Analog-to-Digital... 171

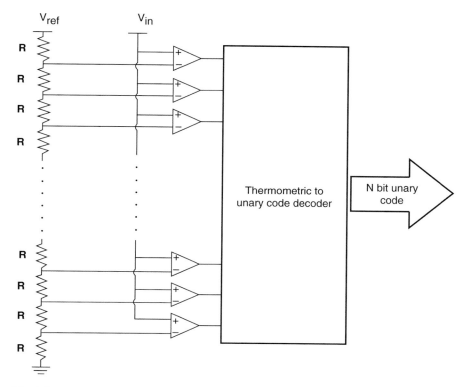

Fig. 4.6 Flash ADC structure

Fig. 4.7 Simplest comparator

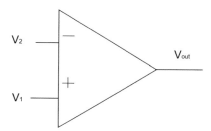

are basically high-gain OpAmps without feedback. The input-output characteristic of the comparator is given in Fig. 4.8 [34].

Offset is the input differential voltage when the comparator output voltage is zero. In the case of an ideal comparator, the offset is equal to zero [34–40]. The accuracy of the flash ADC is practically dependent on matching the resistors and the linearity of the comparators. As already mentioned, the reference voltages are obtained by means of a voltage divider, and the value of the reference voltage generated by every ith resistance can be determined by the formula (4.3) [21]:

Fig. 4.8 Offset error

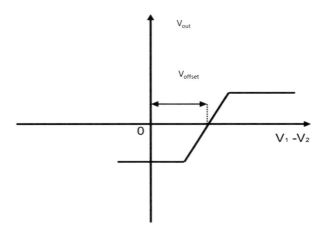

$$V_i = V_{i,\text{ideal}} + \frac{V_{\text{ref.}}}{2^n} \sum_{k=1}^{i} \frac{\Delta R_k}{R}. \qquad (4.3)$$

An ideal comparator assumes a change in the output voltage level only when the voltage values of the inputs equal each other, but in practice the comparator has an offset error and the condition does not occur. In case of offset, the condition (4.4) needs to be met for switching the output of the comparator to "1" or "0" [21].

$$V_{\text{out}} = 1, \text{ when } V \geq V + V_{\text{offset}},\ V_{\text{out}} = 0, \text{ when } V < V + V_{\text{offset}}. \qquad (4.4)$$

From the expression (4.4), it is possible to obtain the voltage value of the switching point of the i-th comparator (4.5) [35].

$$V_{\text{sp}}, i = V_i + V_{\text{offset}}. \qquad (4.5)$$

The offset causes nonlinearity in the overall flash ADC; as a result, there is a DNL and INL error and, in some cases, even missing code (Fig. 4.9) [9].

Comparators are usually designed so that the offset error is minimal, and even if there is a nonlinearity error, the DNL or INL does not exceed 1 LSB, so that there is no missing code. However, random deviations after the manufacturing process lead to a non-monotonic transfer function due to the offset error and thus missing code. To avoid such a problem, it would be appropriate to have a self-calibration offset system implemented, which would allow to correct the offset error and therefore have a monotonic transfer function, avoiding missing code.

Thus, the operating parameters of flash ADCs are inevitably degraded after the manufacturing process. Basically, the nonlinearity of the transfer function in flash ADCs is caused by the deviation of the values of series resistors of reference voltages from the nominal, as well as the nonlinearity of comparators. The reasons for nonlinearity of comparators are offset and amplification errors.

4.1 General Issues to Improve Linearity of Signal's Analog-to-Digital... 173

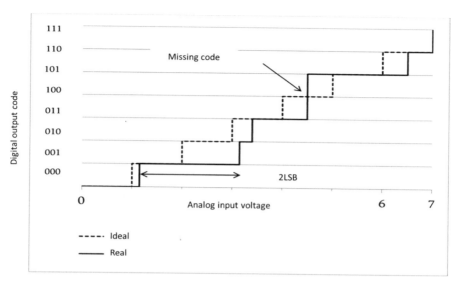

Fig. 4.9 Missing code phenomena in ADC

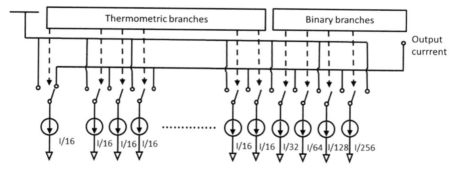

Fig. 4.10 Current DAC

4.1.4 Causes of Nonlinearity Occurrence Signal's in Analog-to-Digital Conversion and the Importance and Necessity of its Reduction in Current DACs

The system embedded in the observed circuit can also improve the operating parameters of the degraded circuit by means of self-calibration. The principle circuit of the current DAC is presented in Fig. 4.10 [41–46]. It consists of current sources of different weights which are connected to the output by a controlled code and provide the appropriate current at the output.

Transistors with metal-oxide-semiconductor (MOS) structure serve as a current source. The current of thermometric branches is responsible for the linear increase of

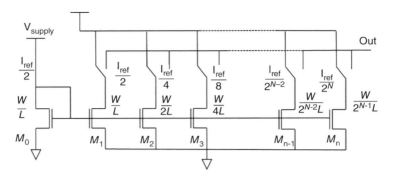

Fig. 4.11 Structure of binary weighted branches

the current, and current sources with binary weight change the current by corresponding values of two degrees (Fig. 4.11) [41].

The branches consist of current mirrors, and the size of the current is determined by the size of transistors (4.6) [41].

$$\frac{I_{dsn}}{I_{ds1}} = \frac{(V_{gs1} - V_{th})^2 \times (1 + \lambda \times V_{ds1})}{(V_{gsn} - V_{th})^2 \times (1 + \lambda \times V_{dsn})}, \quad (4.6)$$

where I_{ds} is the source-drain current of the n-th transistor, V_{gsn} is the gate-source voltage of the n-th transistor, V_{th} is the threshold voltage of the transistor, and λ is the channel length modulation coefficient. The formula is for an ideal case when the parameters and currents of transistors have no deviations.

As in the case of other circuits, there are deviations in the physical parameters of transistors during the manufacturing process. Physical parameters such as oxide thickness, threshold voltage, transistor channel width, and general IC area changes lead to circuit performance degradation [41]. Especially in the case when proper equalization of transistors is not done during the physical design stage of IC, as a result, the sizes of transistors differ from the values set at the beginning of the design, and therefore, the current deviates from the nominal values, and the output current has an incorrect value. Basically, there are three types of deviations: conductance (B), threshold voltage (V), and channel length modulation coefficient (λ) [41]. Deviations can be presented in the following form (4.7):

$$\begin{aligned} B_1 &= B_0 + \Delta B, \\ V_{th \cdot 1} &= V_{th0} + \Delta V, \\ \lambda_1 &= \lambda_0 + \Delta \lambda, \end{aligned} \quad (4.7)$$

4.1 General Issues to Improve Linearity of Signal's Analog-to-Digital...

where ΔB, ΔV, and $\Delta \lambda$ are the deviation sizes of conductivity, threshold voltage, and channel length modulation coefficient. In the saturation mode of M0 transistor, the output current will be represented by the formula (4.8) [41]:

$$I_{ds0} = \frac{B_0}{2} \times (V_{ga} - V_{th \cdot 0})^2 \times (1 + \lambda_0 \times V_{ds0}). \tag{4.8}$$

Taking all deviations into account, the output current in the saturation mode of M1 transistor will be represented by the formula (4.9) [41].

$$I_{ds0} = \frac{B_0 + \Delta B}{2} \times (V_{gs} - (V_{th0} + \Delta V))^2 \times (1 + (\lambda_0 + \Delta \lambda) \times V_{ds0}). \tag{4.9}$$

Therefore, the current of the n-th transistor will be determined by the formula (4.10) [41]:

$$\frac{I_{dsn}}{I_{ds1}} = \frac{B_n}{B_1} \times \frac{(V_{gs1} - V_{th})^2 \times (1 + \lambda \times V_{ds1})}{(V_{gsn} - V_{th})^2 \times (1 + \lambda \times V_{dsn})}. \tag{4.10}$$

Thus, the deviation of the current source from the nominal can be represented as a parallel current source connected to the main current source (Fig. 4.12) [41].

The output current of a non-ideal DAC will be expressed by the formula (4.11) [41]:

$$I_{out}(n) = \sum_{k=0}^{n-1} b_k(n) * I_{out,k}, \tag{4.11}$$

where

$$I_{out,k} = I_{unit} * 2^K + I_{dev \cdot n}. \tag{4.12}$$

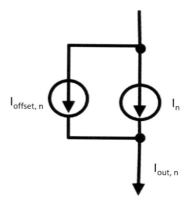

Fig. 4.12 Current source model with current deviation

Such a deviation current will lead to nonlinearity of the current DAC; thus, reducing the self-calibration deviation current in DACs of similar current will lead to an improvement in the linearity of the overall DAC.

Thus, the main cause of nonlinearity in current DACs is the deviations of the current values of the current sources from the nominal value. The main causes of INL and DNL are current sources with large weight—thermometric branches of the current.

4.1.5 Causes of Nonlinearity Occurrence in Signal's Analog-to-Digital Conversion and the Importance and Necessity of Its Reduction in Pipeline ADCs

Another very popular and widely used ADC in modern ICs is the pipeline ADC. It is mainly used in such functions, when the operating frequencies do not exceed 100 MHz. This ADC is slower compared to a flash ADC, but it has a clear advantage in terms of bits. In case of the same area its bit is greater. The simplest pipeline ADC consists of N cascades, each of which receives B bit. In order to obtain a B-bit signal, the input analog signal in each cascade is first selected and stored by a sample-hold (S/H) device, then exposed to coarse quantization (digitization) by means of a sub-range ADC (Fig. 4.13) [47]. The analog voltage is then restored by a sub-DAC, after which the quantized signal is subtracted from the input signal to yield the quantization error. To bring the quantization error to full-scale voltage range, the error is multiplied by 2^{B-1} which is performed by OpAmp (Fig. 4.14) [47–50]. The obtained voltage is applied to the input of the next cascade, which has exactly the same structure. Timing and digital correction is done for the obtained B

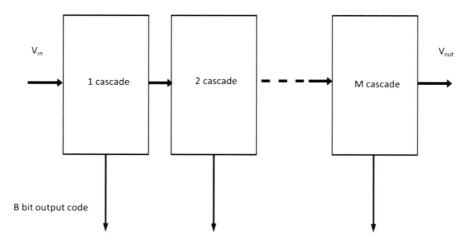

Fig. 4.13 Pipeline ADC block diagram

4.1 General Issues to Improve Linearity of Signal's Analog-to-Digital...

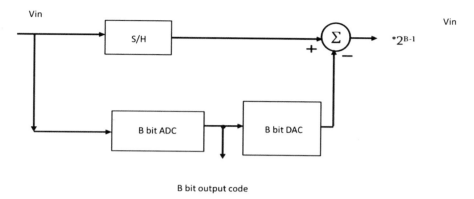

Fig. 4.14 Structure of one cascade

bits by a synchronization and digital error correction device, after which a Y-order digital output corresponding to the input analog signal is obtained.

A few important features can be noted here. The larger B is, the smaller N is, that is, with a smaller number of cascades, the presented operation can be performed. A large choice of B will reduce the size and power consumption of ADC, but the requirements for inter-cascade devices will become stricter [51–54].

It should be noted that if B is chosen equal to the ADC bits, there will be a flash ADC whose bit is not so large. Another feature is that by using the well-known synchronization and digital error correction method, the requirements on the sub-ADCs are significantly eased, allowing the use of comparators with low accuracy and power. On the other hand, since there is a S/H device in each cascade, it is necessary to have an OpAmp with a fast output assertion, which will lead to an increase in power and size, and on which the performance of the ADC will mainly depend. So, for example, a 12-bit pipeline ADC will look as shown in Fig. 4.15 [55–59].

The reason for the inaccuracy can be the deviations of the S/H device and the offset error of the OpAmp.

Low-bit ADCs and DACs have some nonlinearity and other deviations, too. Inaccuracies and errors are presented in Table 4.1 [60]. Inaccuracies, such as the nonlinearity of ADCs and the offset of OpAmps, can be corrected by digital error correction means and offset error compensation methods, respectively [61]. Thus, the above-mentioned inaccuracies are mainly causes of general ADC linearity error. To see the effects of nonlinearity, consider an ADC cascade that contains a 2-bit ADC and assume that the bits of all other cascades are infinitely large. Figs. 4.16 and 4.17 show the I/O characteristics of an ideal 2-bit ADC [60, 62–65]. The switching points are determined by the ADC subconverter, and the switching magnitude depends on the DAC and the OpAmp gain. The switching points of ADC are 0 and $\pm 1/2 V_{ref.}$ and their respective DAC output levels $\pm 1/4 V_{ref.}$ and $\pm 3/4 V_{ref.}$

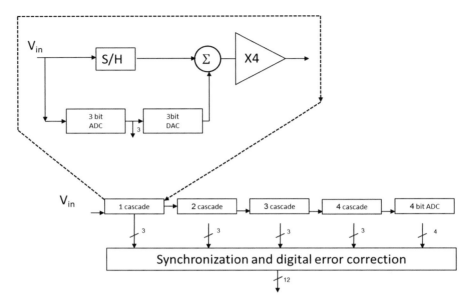

Fig. 4.15 12-bit pipeline ADC

Table 4.1 Error sources and resulting inaccuracies in ADC

Source of error	Occurring problems
Offset error	DC voltage offset
S/H nonlinearity	Nonlinearity
Gain error	Missing code
Offset error	Nonlinearity
ADC nonlinearity	Missing code
Gain error	
Offset error	DC voltage offset
DAC nonlinearity	Nonlinearity
Gain error	Missing code
Offset error	DC voltage offset
OpAmp nonlinearity	Nonlinearity
Gain error	Missing code

The gain of the amplifier is equal to 4. Nonlinearity causes inaccuracy in ADCs, particularly by affecting the magnitude in the case of different codes.

Due to the inaccuracy, the output does not correspond to the switching range of the next cascade, so the overall ADC has nonlinearity, and there are positive and negative DNLs in the transfer function (Figs. 4.18 and 4.19) [60].

OpAmp gain error causes the output to be greater or less than the switching range of the next cascade, depending on whether the gain of the amplifier is greater or less than the ideal gain [60].

4.1 General Issues to Improve Linearity of Signal's Analog-to-Digital... 179

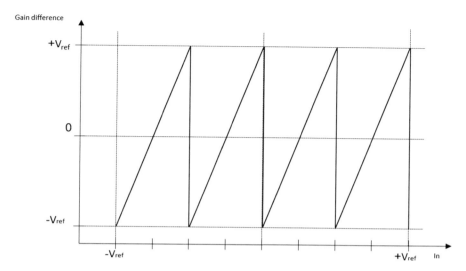

Fig. 4.16 Input-output characteristic of an ideal 2-bit ADC

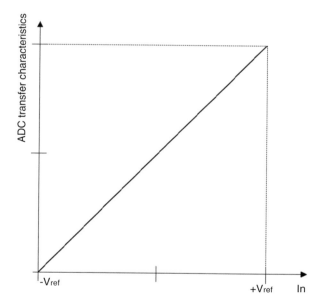

Fig. 4.17 ADC transfer characteristic

Thus, gain error leads to a negative DNL if the gain of the amplifier is less than the ideal (Figs. 4.20 and 4.21), and a positive DNL if the gain is greater than the gain of the ideal amplifier (Figs. 4.22 and 4.23) [60].

Thus, OpAmps are the main source of nonlinearity in pipeline ADCs. Offset or gain errors in OpAmps cause DNL and INL of the entire ADC to increase. All

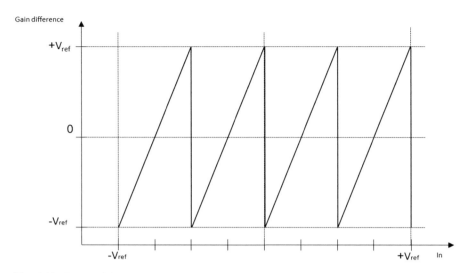

Fig. 4.18 Errors arising due to nonlinearity

Fig. 4.19 Transfer characteristic of an ADC with nonlinearity

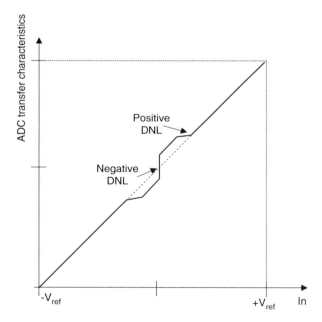

discussed ADCs and DACs inevitably suffer from performance degradation, and it is clearly necessary to have solutions that will significantly improve the performance of the presented converters, allowing to have uninterrupted operating IC.

4.1 General Issues to Improve Linearity of Signal's Analog-to-Digital... 181

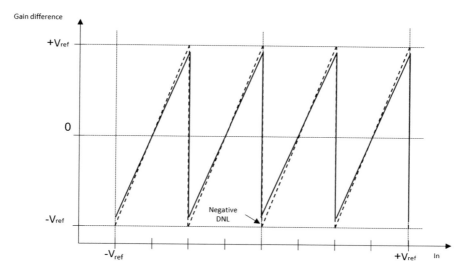

Fig. 4.20 A negative DNL occurring due to small gain

Fig. 4.21 Transfer characteristic of ADC with negative DNL

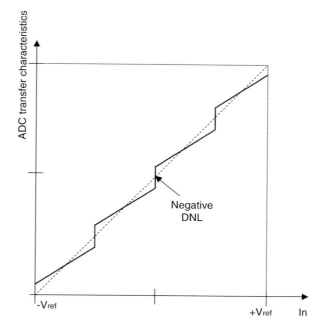

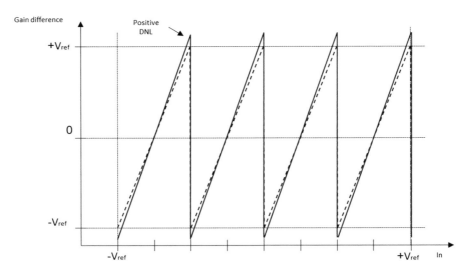

Fig. 4.22 A positive DNL occurring due to large gain

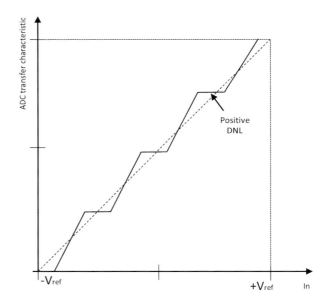

Fig. 4.23 Transfer characteristic of ADC with positive DNL

4.1.6 Existing Means to Improve the Linearity of Signal's Analog-to-Digital Conversion with Self-Calibration

Many methods currently exist to reduce the nonlinearity that occurs in all presented ADCs and DACs. The application of methods depends on the specific type of ADC and DAC and, in particular, on the function they will perform in IC. The efficiency of the means and the accuracy of the nonlinearity reduction are also chosen based on the specific function. The main parameters are the area occupied on the total IC, power consumption, as well as the financial limitations.

Such means in turn have inaccuracies and may cause additional errors. Therefore, the proposed solutions should be as accurate as possible and as linear as possible. It is also very important that self-calibration of the system is performed in the sequence of minimum steps, because the accuracy of the system also depends on the number of steps: the smaller the sequence of steps, the less the total effect of the errors made in each step.

The nonlinearity reduction is done by feedback through embedded systems, the purpose of which is to measure the nonlinearity error and reduce the nonlinearity error with the intended accuracy through a specific algorithm. Deviated parameters (current, offset, parameters of electrical elements) are corrected by auxiliary devices: calibration DACs, additional current sources, and preamplifiers. Typically, the testing and measurement algorithm is implemented using a finite state machine (FSM). In many cases, the FSM also performs a calibration function.

4.1.7 Existing Methods for Reducing Nonlinearity in Flash ADCs

As mentioned, the flash ADC is the fastest ADC and contains a large number of comparators that insert offset error, resulting in deterioration of linearity. One of the proposed solutions to increase the linearity of this ADC is observed. The presented flash ADC (Fig. 4.24) consists of a track and hold amplifier (THA), a reference voltage generator matrix, four-cascade preamplifiers, comparators, an encoder, a calibration feedback circuit, and an output cascade. Cascaded preamplifiers are designed to increase the difference between the reference voltage and the input voltage to overcome the offset error of comparators. Interpolation allows to generate intermediate points of intersection with zero. Therefore, it is not necessary to compare the input with the 255 reference voltages; at the same time, the number of preamplifiers decreases, saving a large area. The first two cascades generate 33 zero-crossing points, which are corrected by the calibrating circuit. The interpolation performed by means of resistors is performed in the second and third cascades, as a result of which the remaining points of intersection with zero are obtained [66].

A pseudo-differential no-feedback THA (Fig. 4.25) is introduced to prevent the signal-to-noise ratio from increasing [66]. The pseudo-differential p-type MOS (p-MOS) source repeater is usually used as a repeater in succeeding preamplifiers because it has high linearity and good offset characteristics [66–70].

However, due to the short channel, there are problems such as the reduction of the output voltage range of ADC, and in order to have the same accuracy, it is necessary to increase the power consumption.

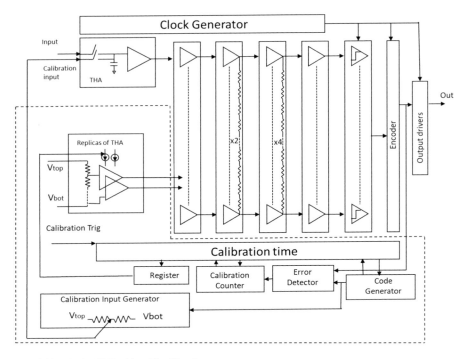

Fig. 4.24 Flash ADC with self-calibration system

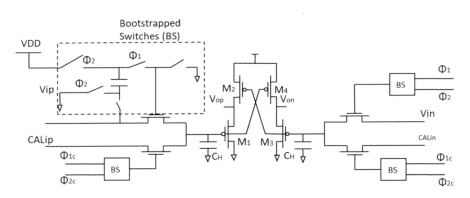

Fig. 4.25 Pseudo-differential THA without feedback

Cascading current sources and the use of long-channel transistors can improve the output voltage range, resulting in increased linearity, but they may require larger voltage drops and also affect speed [70–74]. To soften this drawback, the gates of M2 and M4 transistors are connected together instead of connecting the input signal to the reference voltage (Fig. 4.25). Here, M2 and M4 operate as dynamic current sources for M1 and M3 source repeaters [66]. As the input Vip increases, ID1 decreases due to the decrease of source-drain voltage of M2 transistor. However, the complementary Vin input decreases due to ID1 current. This approach increases the gain of the amplifier. A similar structure provides a parallel opportunity to improve signal rise/fall times without consuming static power. Thus, the gain, gain bandwidth product, and the slew rate are improved by 20, 40, and 30%, respectively [66].

In CMOS flash ADCs, circuit element mismatches and manufacturing inaccuracies cause offset errors in comparators and preamplifiers. Averaging methods can partially solve the problem and increase the linearity of ADC; however, long-channel transistors are still required in preamplifiers, and degradation of linearity is inevitable. To solve this problem, a calibration circuit is proposed, which calibrates the reference voltage by means of correction currents in such a way that offset errors caused by preamplifiers and comparators are neutralized. The part marked with dotted lines in Fig. 4.24 is responsible for the self-calibration process. The digital code generator generates codes corresponding to the ideal ADC output, which are transferred to calibration input generator. The calibration input generator contains precision-valued resistors and can provide linearity equal to 10-bit. Φ1c and Φ2c keep cal.ip and cal.in when Φ1 and Φ2 turn off Vip and Vin (Fig. 4.25). The ADC quantizes the calibration input instead of the actual input. The correction currents are then determined by comparing the output of the ADC with the ideal code based on the amount of error obtained. The range of current correction is determined by the bit rate of the control counter and the discrete current source outside IC. A similar circuit allows nonlinearity correction up to ±10 LSB with a correction step of 0.2 LSB (Fig. 4.26) [66]. After calibration, the calibration inputs are switched to ADC inputs, and the circuit returns from calibration mode to operating mode. The maximum calibration time is 0.12 ms [66].

The offsets are also calibrated as a result of comparison. The offset errors of the third and fourth cascade preamplifiers and comparators are not corrected. The circuit was run for 24 h, after which the deviations of the parameters were checked, which did not undergo significant changes. In order to test the proposed circuit, a sinusoid with a frequency of 2.3 MHz and a holding speed of 1.25 GS/s was taken as an input analog signal. Without self-calibration, the INL of ADC was measured, and in the worst case it is equal to 3.3 LSB. After the self-calibration mode, the INL of the circuit is reduced up to 1.1 LSB, and the DNL is reduced to 1.3 LSB [66].

Thus, the system of linearity increase with self-calibration in flash ADC partially solves the problem of nonlinearity reduction, because the values of INL and DNL are still greater than 1 LSB, and therefore, missing loss will inevitably occur, which is unacceptable. In addition, the proposed solution occupies a large area, and the self-calibration algorithm consists of a large number of successive steps, as a result of which additional errors in each successive step lead to a large error in total.

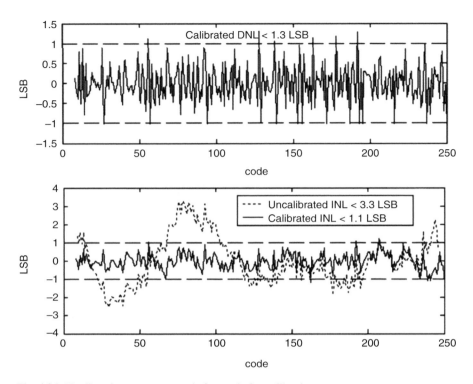

Fig. 4.26 Nonlinearity measurements before and after calibration

4.1.8 Existing Methods for Reducing Nonlinearity in Current DACs

The main cause of nonlinearity in current DACs is the mismatch of current source transistors to each other, which leads to deviation of currents from nominal values. Current deviation is a static error, so it is corrected once when the circuit is switched on. There are many approaches to solving the problem. The most general view of the solution is presented (Fig. 4.27) [5, 75–80]. Current deviation error correction methods are based on three main operations: self-checking or self-error measurement, providing a correction process with a feedback circuit or algorithm, and the self-calibration operation implemented by it. The deviation current correction range of each tuning DAC (reg DAC) is determined by the maximum possible thermal current deviation, and the LSB size of the tuning DAC determines the tuning accuracy. Note that the LSB step size should be small enough to ensure the necessary linearity. With such DACs, it is possible to correct both negative and positive current deviations. A 1-bit ADC is used to find the current error, which itself is already linear. Only the input deviation error of the current is inevitable, I_{offset} [5, 76].

4.1 General Issues to Improve Linearity of Signal's Analog-to-Digital...

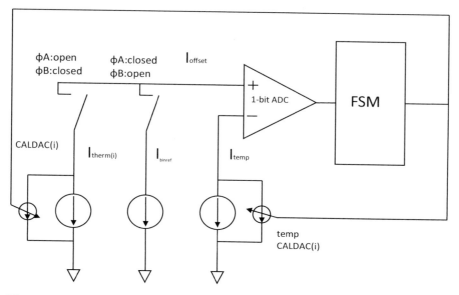

Fig. 4.27 Linearity self-calibration system in current DAC

The input deviation error is corrected in two stages (Fig. 4.28) [5, 76]. During phase A (φA), the temporary current source, I_{temp}, is calibrated according to the reference current, I_{binref}.

Ioffset is also held; during phase B the reference current is applied to the input of the comparator to calibrate I_{therm}. I_{therm} is calibrated according to the temporary current I_{temp} while correcting the input deviation current of the comparator, I_{offset} (4.13).

$$\varphi A : I_{temp.} \approx I_{binref} - I_{offset}, \varphi B : I_{therm.} \approx I_{temp.} + I_{offset} \approx I_{binref} \quad (4.13)$$

During the two stages of calibration, Ioffset is eliminated. The quantization error occurring during each cycle is controlled to reduce. For I_{temp} to be always set according to positive quantization error of I_{binref}, a polarity check is performed after state V3, while a check is performed after state V6 to make sure that $I_{therm(i)}$ is calibrated according to I_{temp}, but already with a negative quantization error.

Such a sequence of operations allows to control the distribution of quantization error and reduce the post-calibration error [5]. In calibration process, the total current of the binary weighted current sources, plus an empty current source with one LSB weight, is selected as the reference current source for comparison. This approach practically eliminates the DNL error when converting from binary to thermometric codes. Then calibration is done only for thermometric branches of the current, since their effect on the linearity of the DAC is much more significant compared to the binary branch currents. INL was measured before and after using the method. Before self-calibration, INL of the circuit >1.3 LSB, after calibration INL < 0.4 LSB

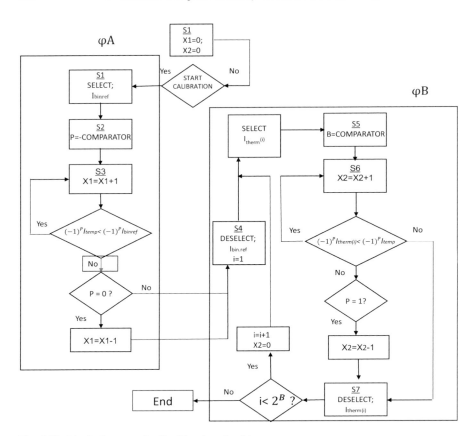

Fig. 4.28 Block diagram of self-calibration algorithm

(Fig. 4.29) [5]. The reason for the remaining nonlinearity is the non-calibrating binary part of the circuit.

Thus, the presented system of self-calibration of the linearity of the current DAC shows significant results; in particular, it reduces the DNL and INL of the system to the extent that the missing codes in the system are absent. The nonlinearity error suppression range of the proposed number is rather small, and the system is effective in suppressing small linearity errors and therefore does not completely solve the problems presented.

4.1 General Issues to Improve Linearity of Signal's Analog-to-Digital...

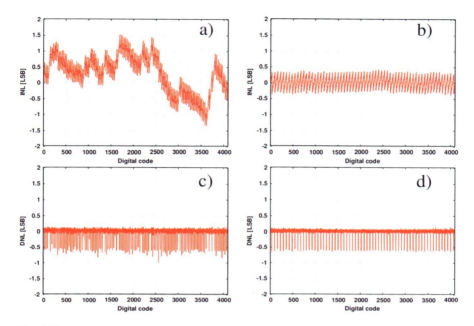

Fig. 4.29 Nonlinearity measurements before (**a**, **c**) and after (**b**, **d**) calibration

4.1.9 Existing Methods for Reducing Nonlinearity in Pipeline ADCs

A high gain OpAmp with negative feedback allows for precise operation by reducing the nonlinearity present in non-feedback circuits. The gain of a simple feedback circuit can be calculated using the following formula:

$$A_{cl} = \frac{A}{1+A\beta} = \frac{1}{\beta}\left(\frac{1}{1+1/A\beta}\right), \qquad (4.14)$$

where β is the feedback coefficient and A is the gain of the OpAmp. For an ideal OpAmp, when the gain is infinitely large, A_{cl} is equal to $1/\beta$ [81]. However, A_{cl} is smaller compared to $1/\beta$. If β is reduced, A_{cl} can be equalized to the desired value. In pipeline ADCs, the residual voltage is amplified by feedback multiplier DACs (MDACs). The gain error caused by the small gain of the OpAmp can be corrected on the account of feedback coefficient [81–90].

Figure 4.30 shows an MDAC with a calibration capacitor (C_{cal}), the size of the capacitance of which can be controlled by digital means. During the holding phase, eight capacitors (C_h) control the input. Both C_{cal} and feedback capacitance (C_f) are discharged, connecting to the DC voltage. During the amplification stage, the sub-DAC's digital output C_f is connected to either $+V_{ref}$ or $-V_{ref}$. Through the error overlap method, C_f gets equal to $2C_s$; therefore, it has an inter-cascade gain

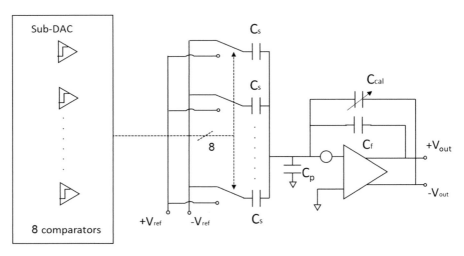

Fig. 4.30 MDAC with calibration capacitor

equal to 4. Ch is enabled as a negative feedback capability and C as a positive feedback capability. Thus, the amplification of MDAC is given by the following formula:

$$V_{out} = A_v \left(V_{in} - \sum_{i=1}^{8} D_i \frac{V_{ref.}}{8} + \frac{8C_s + C_f + C_{cal} + C_p}{8C_s} V_{offset} \right), D_i = \pm 1, \quad (4.15)$$

where Cp is parasitic capacitance and Voffset is the offset of OpAmp, and Av is inter-cascade gain of MDAC and is expressed by:

$$A_v = \frac{8C_s}{C_f + \frac{8C_s + C_f + C_{cal} + C_p}{A} - C_{cal}} \quad (4.16)$$

It is desirable that A_v is equal to 4. By controlling C_{cal}, gain equal to the desired value of 4 can be obtained. The self-calibration functional diagram of pipeline ADC is shown in Fig. 4.31 [81].

The OpAmp of the last cascade MADC is assumed to have sufficient gain to meet the sub-ADC requirements and provide sufficient accuracy. Therefore, it does not need to be calibrated, it is assumed to be ideal.

The calibration process starts from the penultimate MADC. 1/8 V_{ref} voltage is injected into the calibrating MADC during the hold phase.

During the amplification stage, the feedback MADC amplifies 1/8 V_{ref} by A_v times, and the ideal MADC holds the output of the calibrated MADC, then the ideal MADC subtracts 3/8 V_{ref} voltage and also amplifies it four times. If the gain of the MADC in calibration stage is 4, then the output voltage of the ideal MADC should be equal to 1/2 V_{ref}. The output of the ideal MADC is compared to 1/2 V_{ref} voltage

4.1 General Issues to Improve Linearity of Signal's Analog-to-Digital...

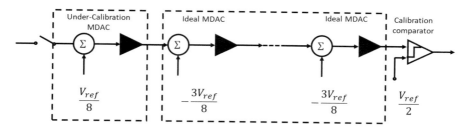

Fig. 4.31 Functional diagram of pipeline ADC self-calibration

using a calibration comparator. The output of the comparator is then stored in a sequential approximation register which digitally controls C_{cal} capacitor size. After a few cycles, the calibration information stored in the register will set the gain of the MADC in calibration stage to 4. Then the previous MADC appears in the self-calibration phase and so on until all MADCs are calibrated and the gain error is minimized. Fifty-six cycles of the self-calibration process are needed to calibrate three MADCs. The entire self-calibration process takes 168 cycles [81, 91–99]. In Eq. 4.15 $(8C_s + C_f + C_{cal} + C_p) V_{offset}/ 8C_s$ component that occurs due to offset error of the OpAmp does not hinder the operation of the circuit because it is neutralized due to the application of the overlap method. Ideal MADCs work as analog signal converters that amplify the signal and give a stabilized value at the output. The deviation voltage of the OpAmp leads to a deviation of the voltage value at the output, which then serves an input to the other cascades and can lead to an error in self-calibration process. The input deviation error is stored in a register so that the deviation error can be reduced. Analog self-calibration methods usually require accurate reference sources. C_s in the first MADC is designed to have less than 0.1% deviation, and 1/8 V_{ref} input voltage can be generated by the MADC if it is used as a switching capacitor as shown in Fig. 4.32. The inputs of the MADC in calibration stage are connected to $+V_{ref}$, $-V_{ref}$, and Vgnd with 4Cs, 3Cs, and Cs capacitors, respectively; thus, the total input voltage is equal to 1/8 V_{ref} [81, 100–109].

To use the offset hold method, the offset of the OpAmp is held when the amplifier is in feedback. During the amplification phase $+V_{ref}$ and $-Vref$ are connected to the input with $4C_s$ capacitors to get zero voltage (Fig. 4.33). Cf and Ccal are connected to the positive and negative outputs of the OpAmp, respectively. For an ideal MADC, all $8C_s$ capacitors store the initial MADC output in the hold phase (Fig. 4.34). The offset of the OpAmp is also stored. Then, in the amplification phase, $+V_{ref}$, $-V_{ref}$, and Vgrnd are connected to the input with 5Cs, 2Cs, and Cs capacitors, together generating 3/8 V_{ref} voltage (Fig. 4.35). The residual voltage is amplified by a feedback circuit [81].

Before the calibration process, the INL and DNL of the circuit are 1.7/−1.0 LSB and + 15.6/−15.2 LSB, respectively. After calibration, INL and DNL are equal to 0.7/−0.6 LSB and 0.8/−0.9 LSB, respectively (Fig. 4.36) [81].

Fig. 4.32 Calibration through switching capacitors (holding phase)

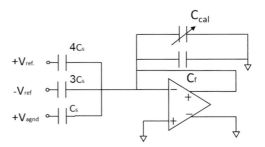

Fig. 4.33 Calibration through switching capacitors (amplification phase)

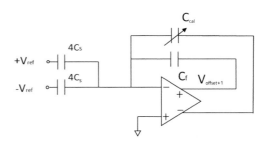

Fig. 4.34 Ideal MADC (holding phase)

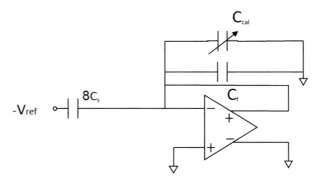

Thus, the presented self-calibration linearization reduction method significantly reduces the nonlinearity of pipeline ADC by reducing the INL and DNL errors to such values that missing loss is not observed experimentally. However, the implementation of the method presupposes the presence of such elements in the system, the linear parameters of which are ideal, and their implementation is practically very difficult, if not impossible. Therefore, the question of developing other means remains relevant.

4.1 General Issues to Improve Linearity of Signal's Analog-to-Digital... 193

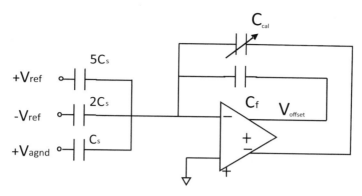

Fig. 4.35 Ideal MADC (amplification phase)

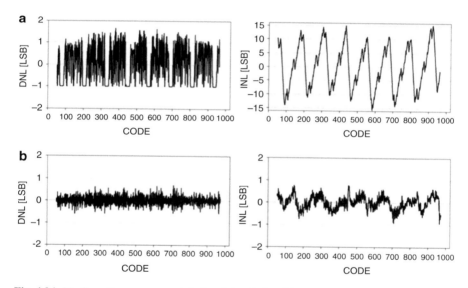

Fig. 4.36 Nonlinearity measurements before (**a**) and after (**b**) calibration

4.1.10 Principles to Improve the Linearity of Signal's Analog-to-Digital Conversion with Self-Calibration

The existing means and approaches for improving the linearity of ADCs and DACs with self-calibration do not fully meet currently formed requirements. Therefore, the development of new means and solutions is relevant. Thus, the following principles are proposed:

1. Introduce feedback in flash ADC, which will reduce the offset error of common comparators by means of auxiliary DACs, FSM, and preamplifier. This will allow to reduce the nonlinearity of the whole system.
2. Add a feedback system to the current DAC, which, by comparing the currents of the current sources with a larger current source deviating less from the nominal one, will allow to calibrate the currents of the latter, thus making it possible to reduce the nonlinearity of the DAC.
3. Introduce an amplifier offset error system in pipeline ADC, which will allow to reduce the nonlinearity of the latter, thus increasing the linearity of ADC.

The mentioned principles will allow to reduce the errors, made in manufacturing process and, therefore, to increase yield percentage of working circuits, reducing manufacturing costs.

4.2 Conclusions

1. Deterioration of operating parameters in digital-to-analog and analog-to-digital converters leads to an unacceptable increase in nonlinearity errors. Taking into account the impossibility of eliminating deviations in the parameters of integrated circuits of the specified class after the end of manufacturing process, the development of embedded self-calibration deviation correction systems has become important at present.
2. The analysis of the existing means of improving the linearity of analog-to-digital conversion by self-calibration shows that, although an attempt was made to eliminate the main causes of offsets and amplification errors, the degree of correction of deviations does not meet modern requirements. For this reason, it became necessary to develop new means of building embedded systems of correction of deviations with self-calibration.
3. Principles of development of means for improving the linearity of the analog-to-digital conversion with self-calibration are proposed, which, unlike the known solutions, are more effective from the point of view of reducing the nonlinearity. Systems with built-in feedback allow to ensure a sufficient reduction of deviated parameters in case of an increase in the area occupied on the die and the calibration time within the permissible limits.

4.3 Methods of Improving the Linearity of Signal's Analog-to-Digital Conversion with Self-Calibration

4.3.1 Method of Reducing the Nonlinearity with Self-Calibration by Correcting the Offset Error of Comparators in Flash Analog-to-Digital Converters

From the discussion in Sect. 4.1 it follows that the existing measures do not completely satisfy and do not solve the existing problems, and even if they solve them partially, they require large-scale resources in terms of area and power.

The self-calibration flash ADC linearity improvement method discussed in Sect. 4.1 required large area and, despite that, did not solve certain problems, such as the preamplifier offset problem, which also leads to linearity degradation. The discussed flash ADC circuit largely solved the accuracy issue of reference voltages by matching them to the expected voltage values, using multiple components in the feedback circuit such as registers, ideal code generators, error detectors, as well as discrete linear elements for matching, such as the debug input generator. To obtain reference voltages and reduce the effect of offset error, a large number of pre-amplifiers are used, which in turn have offset error and therefore increase the nonlinearity of the overall system.

In order to reduce the drawbacks listed above, a method is proposed, for the implementation of which a traditional 8-bit flash ADC is used. A flash ADC uses a voltage divider with linear resistors to obtain reference voltages [110–118]. Then the reference voltages are applied to the inputs of comparators, where they are compared with the input signal, and when the reference voltage coincides with the value of the input signal, there is a corresponding digital signal at the output of the comparator (Fig. 4.37). The code received at the output of the comparators is thermometric; that is why a thermometric-to-unary code converter is connected at the output. Flash ADC contains 2^{n-1} comparators (Fig. 4.38). The comparator is a simple telescopic amplifier with high gain, whose output cascade is a common-source single-cascade amplifier [119]. As it was mentioned in Sect. 4.1, due to the offset error of amplifiers, the nonlinearity in the circuit increases and is the main cause of the nonlinearity. This problem was not completely solved in the past, but the consequences of the error were corrected, which did not give the desired result.

A circuit (Fig. 4.39) was proposed to reduce the offset error of the comparator, the main idea of which is to use one preamplifier at the input of the comparator instead of multiple preamplifiers, which will reduce the input offset error of the comparator through feedback [119].

The offset error correction circuit of the comparator has two operating modes: calibration and operating. In the calibration mode, the inputs of the circuit are connected together by means of a switch. Calibration is done using calibration DACs and a successive approximation algorithm. The algorithm is implemented using FSM [119].

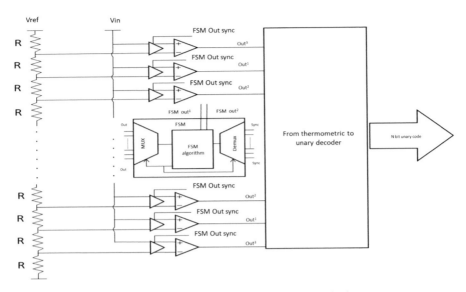

Fig. 4.37 Block diagram of a flash ADC and general view of the method

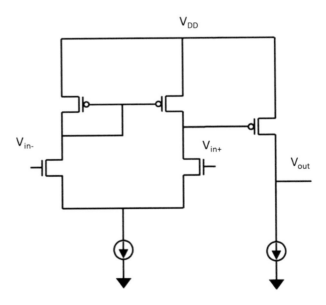

Fig. 4.38 Comparator circuit

Calibration DAC consists of binary weight current sources to which a register is connected (Fig. 4.40). The register allows to save the current values obtained as a result of calibration. During the calibration stage with calibration DACs, the voltage at the output of the preamplifier is established, in the case of which the input offset of the comparator is reduced as much as the calibration DAC allows. Then the circuit is

4.3 Methods of Improving the Linearity of Signal's Analog-to-Digital...

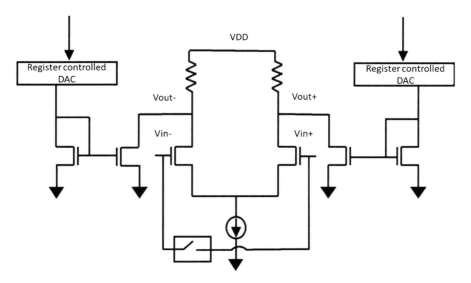

Fig. 4.39 Preamplifier with calibration DACs

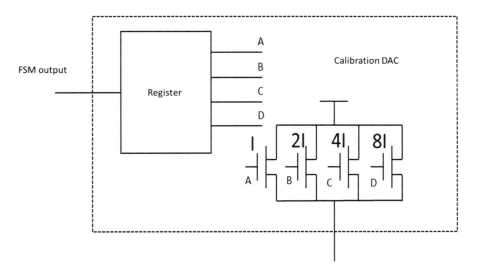

Fig. 4.40 Calibration DAC

brought to the actual operating mode, where the input offset is reduced. The FSM is connected to the next comparator and the same process is performed until the offset error of all comparators is reduced to possible minimum. Thus, the linearity of the overall system increases. Two types of preamplifiers are used because there is a wide voltage range in the system. The presented circuit is used for voltages from Vref to

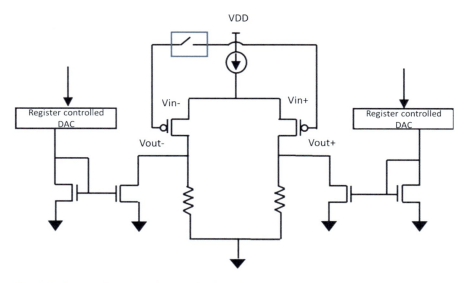

Fig. 4.41 Preamplifier with calibration DACs with P-MOS structure

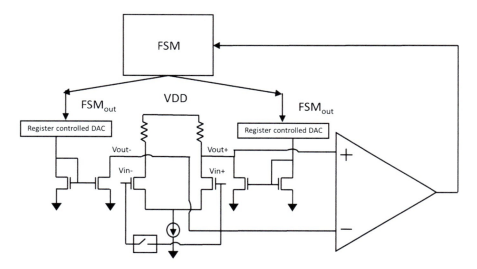

Fig. 4.42 Self-calibration circuit with feedback

Vref/2, and for the remaining domains, the circuit with p-MOS input transistors is used (Fig. 4.41) [119].

In both circuits, the calibration DACs have the same structure. The only difference is the change of the type of transistors of the input pair.

The complete view of the comparator offset correction system and feedback is presented in Fig. 4.42 [119].

4.3 Methods of Improving the Linearity of Signal's Analog-to-Digital... 199

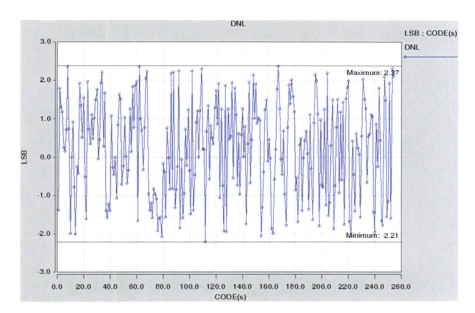

Fig. 4.43 DNL before calibration

In the calibration DAC, the current values of the current sources are set to neutralizing the maximum possible offset value, with an added 10% margin.

INL and DNL measurements were taken before (Figs. 4.43 and 4.44) and after (Figs. 4.45 and 4.46) calibration to evaluate the effectiveness of calibration. INL decreased by about 70% and DNL by 73%. The nonlinearity error is less than 1 LSB; therefore, no missing code was observed [119].

The presented simulation results correspond to the worst case with the largest deviation in the manufacturing process, and the correction of the operating parameters was performed in the specified case. The complete results of DNL and INL are given in Tables 4.2 and 4.3.

The layout of the designed flash ADC is presented in Fig. 4.47.

An external clock signal was used for FSM switching. Correction and equalization of duty cycle of the clock signal to the nominal value of 50% was made, so that it does not cause additional inaccuracies. The correction was made using duty cycle corrector (DCC) (Fig. 4.48) [120].

The DCC consists of a duty cycle detector (DCD), a duty cycle regulator (DCR), and a CML-CMOS buffer. Having received the control signal V_c, the DCR equalizes the constant voltage components of the input signals due to the negative feedback. DCD detects the deviation of CMOS signals from the average value. Then the DCR calibrates the duty cycle of the signal, having at the output a clock signal with a duty cycle close to the nominal, but not fully differential (Fig. 4.49). Then it brings CML-CMOS signal into full differential form. It consists of a differential amplifier, the inputs of which are connected to capacitors to filter the constant component of

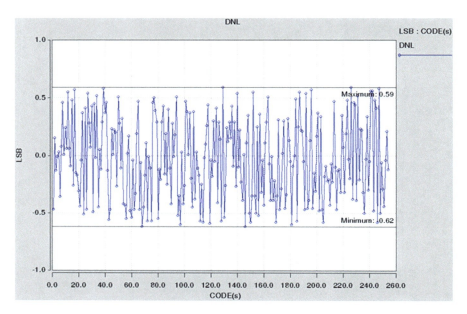

Fig. 4.44 DNL after calibration

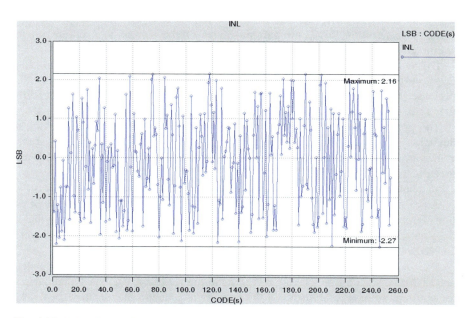

Fig. 4.45 INL before calibration

4.3 Methods of Improving the Linearity of Signal's Analog-to-Digital...

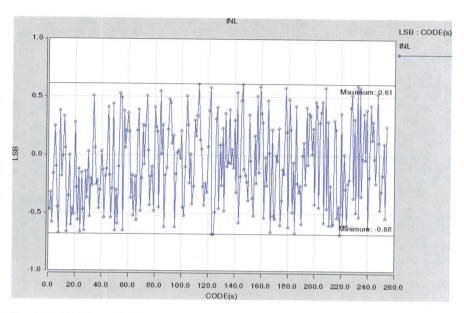

Fig. 4.46 INL after calibration

Table 4.2 Simulation results of the designed flash ADC DNL (LSB)

Temperature/process	Typical	Slow	Fast
−40	0.35	0.60	0.48
25	0.42	0.54	0.37
125	0.51	0.62	0.58

Table 4.3 Simulation results of the designed flash ADC INL(LSB)

Temperature/process	Typical	Slow	Fast
−40	0.45	0.63	0.49
25	0.36	0.54	0.32
125	0.42	0.68	0.52

the signal, two n-MOS transistors, which are necessary to equalize the constant components of the voltage at the input to the CML-CMOS buffer. The CML-CMOS buffer (Fig. 4.50) is necessary to obtain a fully differential signal [120].

One of the constituent parts of the DCD is the differential amplifier, which ensures the stability of the circuit with feedback, the high operating frequency, and the large-scale duty cycle correction layer. A schematic view of the DCD is shown in Fig. 4.51 [120].

As shown in Fig. 4.51, DCD contains a coupled circuit that improves the accuracy and performance of the circuit. Then the differential amplifier and filters provide the control voltages. The control voltages are connected to the DCR by feedback and calibrate the duty cycle until it is as close to the nominal value as possible [120].

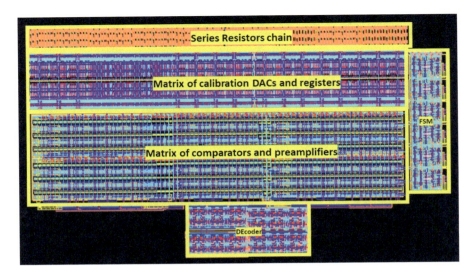

Fig. 4.47 Layout of flash ADC

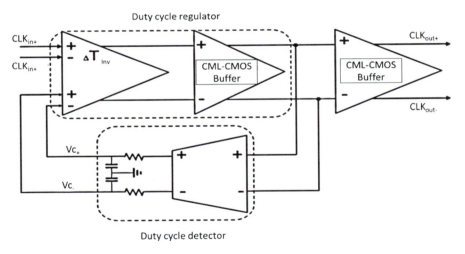

Fig. 4.48 Block diagram of designed analog DCC

Thus, duty cycle, close to the nominal, is obtained, and therefore, the clock signal does not cause an error in the operation of the circuit. A neg-C circuit was used to reduce parasitic capacitances and their effects [121].

Figures 4.52 and 4.53 show power supply rejection ratio (PSRR) and common mode rejection ratio (SMRR), respectively [120].

A neg-C circuit was used to reduce parasitic capacitances and their effects [121].

Thus, the built-in system of increasing linearity with self-calibration of flash ADC was developed. Its implementation is based on a feedback system, which reduces the

4.3 Methods of Improving the Linearity of Signal's Analog-to-Digital... 203

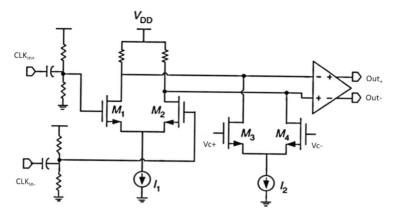

Fig. 4.49 DCR schematic view

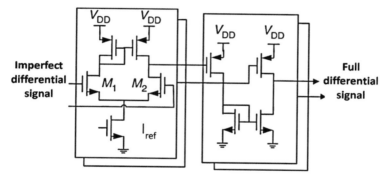

Fig. 4.50 Schematic view of CML-CMOS buffer

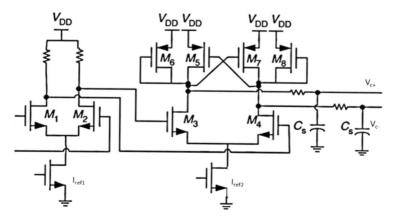

Fig. 4.51 Schematic view of *DCD*

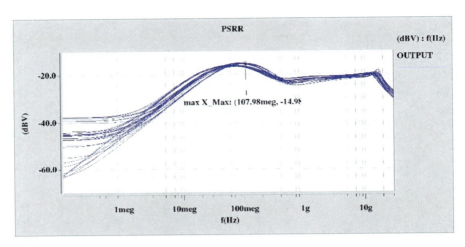

Fig. 4.52 Power supply rejection ratio (PSRR)

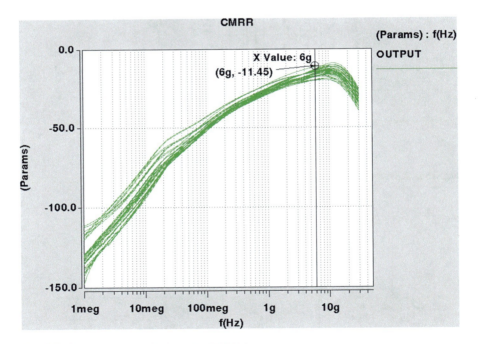

Fig. 4.53 Common mode rejection ratio (CMRR)

nonlinearity of comparators of flash ADC by means of FSM-implemented approximation algorithm and auxiliary current DACs, at the expense of reducing the offset error. Increasing the linearity of comparators leads to increasing the linearity of the overall ADC. The resulting system contains fewer auxiliary elements compared to previous solutions. It reduces nonlinearity parameters by approximately three times, providing DNL and INL less than 1 LSB, due to which no missing loss is observed. Compared to previous approaches, the efficiency of the system is higher. The proposed method has 12% higher calibration time.

4.3.2 Method of Reducing the Nonlinearity with Self-Calibration by Correcting the Current Deviation Error of Current Sources in Current Digital-to-Analog Converters

As discussed in Sect. 4.1, current source mismatches in current DACs cause the DAC transfer function to be nonlinear. In order to solve the problem, in most of the existing circuits, a comparison is made with the calibrated current and the total currents of all the binary currents. As a result of comparison, the value of the calibrated thermometric current source is equalized to the total current of binary weighted current sources. The idea of this method is that the currents of all thermometric branches are equal to the nominal value, and the transfer function is more monotonous in the thermometric part. The basis of the method is the idea that the deviations of separate binary weight current sources in different directions lead to the fact that the value of the total current is close to the nominal, ideal value of the thermometric current. But the circuit has a drawback [17]. In case that the deviations of binary weighted transistors are in one direction, the current deviation is also in one direction, and therefore a deviation from nominal occurs. A method was proposed to avoid the problem (Fig. 4.54).

It is known that the larger physical dimensions of the electrical component, the smaller the deviation of its electrical parameters from the nominal. Taking into account the above, it is possible to use a large-scale transistor as a reference current source, the current of which will be equal to the nominal value, and it will itself be compared with the currents of thermometric branches, and according to its current, the correction of the current of the thermometric branches will be made. Monte Carlo simulations were performed to evaluate the current deviations of binary weighted transistors and a single large current source (Figs. 4.55 and 4.56) [17].

The current of the large current source deviates from the nominal by ±10%, and the total current of the binary weight transistors—±25%; therefore, in the process of

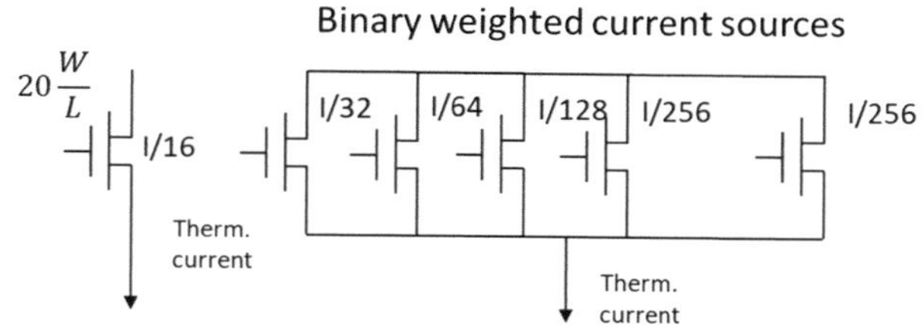

Fig. 4.54 Characteristic of the method

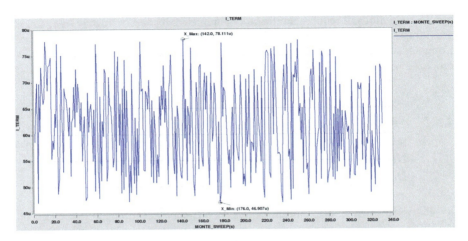

Fig. 4.55 Total current deviation of binary weighted current sources

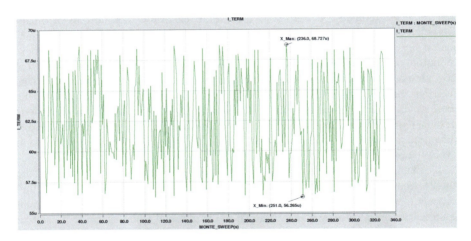

Fig. 4.56 Total current deviation of large current sources

calibrating the thermometric current sources, a large-scale transistor should be used as a reference current, and not the total current of binary weight transistors. There will be an increase in linearity, in particular an improvement of DNL and INL in the thermometric part. Compared to the previous method, the INL error will be smaller, because the current of thermometric branches will be less deviated from the nominal value, which follows from the presented results [17].

The comparison and equalization between currents is performed using a 1-bit ADC and FSM, in which the FSM algorithm represents a successive approximation algorithm. A 1-bit ADC is itself already linear and therefore is not a source of nonlinearity. In addition, to calibrate the weights of the currents, current calibration DACs are connected to thermometric branches, which, being in feedback with the

4.3 Methods of Improving the Linearity of Signal's Analog-to-Digital...

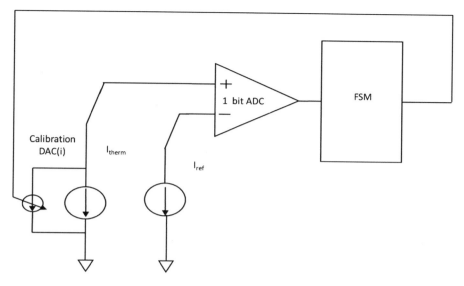

Fig. 4.57 Self-calibration circuit with feedback

Table 4.4 Simulation results of the designed current DAC DNL (LSB)

Temperature/process	Typical	Slow	Fast
−40	0.24	0.29	0.18
25	0.2	0.24	0.23
125	0.22	0.25	0.28

Table 4.5 Simulation results of the designed current DAC INL (LSB)

Temperature/process	Typical	Slow	Fast
−40	0.28	0.39	0.4
25	0.22	0.35	0.33
125	0.31	0.38	0.36

FSM, equalize the currents of thermometric branches to the current value of the nominal current source (Fig. 4.57).

To evaluate the effectiveness of the method, a comparison was made with the previous method. Nonlinearity parameters INL and DNL were measured (Tables 4.4 and 4.5). The results and the layout of DAC are shown in Figs. 4.58, 4.59, 4.60, 4.61, and 4.62 [17].

Thus, an embedded means of increasing the linearity of the current DAC was proposed. The embedded system compares the values of the current sources of the current DAC with the nominal current and performs current matching through feedback. As a result of current matching, the linearity of the current DAC increases. The comparison was made with a large current source, the current of which does not

208 4 Methods to Improve Linearity of Signal's Analog-to-Digital Conversion...

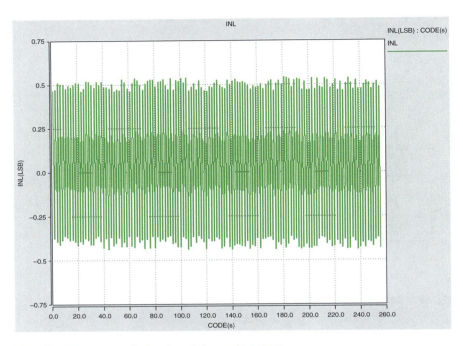

Fig. 4.58 Effectiveness of using the existing method (INL)

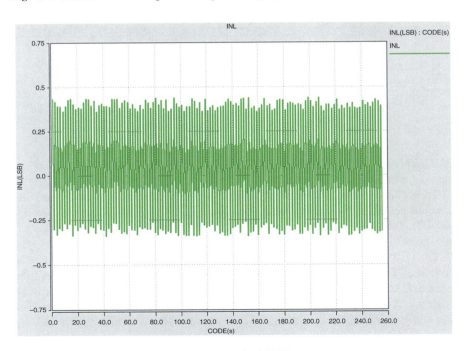

Fig. 4.59 Effectiveness of using the proposed method (INL)

4.3 Methods of Improving the Linearity of Signal's Analog-to-Digital... 209

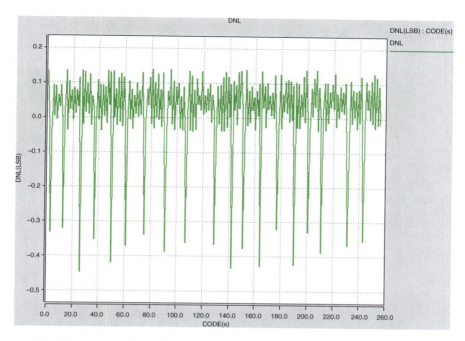

Fig. 4.60 Effectiveness of using the existing method (DNL)

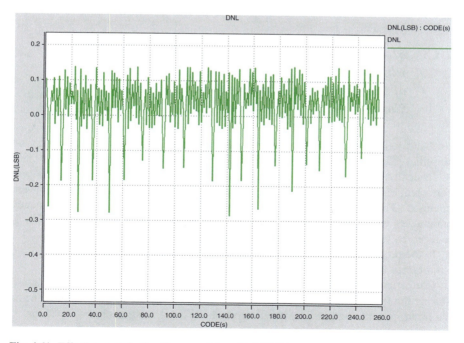

Fig. 4.61 Effectiveness of using the proposed method (DNL)

Fig. 4.62 Layout of current DAC

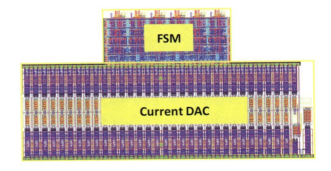

Fig. 4.63 Offset error correction circuit

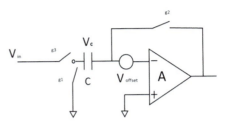

deviate during the manufacturing process. The used area has slightly increased by approximately 5% compared to existing solutions, resulting in a 20–25% more efficient system.

4.3.3 Means of Reducing System Nonlinearity with Self-Calibration by Increasing the Linearity of Comparators in Pipeline Analog-to-Digital Converters

As discussed in Sect. 4.1, the three main causes of nonlinearity in pipeline ADCs are amplifier gain error, amplifier offset error, and nonlinearity in sub-DACs or ADCs.

Basically, in the previous works the amplification error of the amplifiers was corrected, but the other two were neglected; however, quite significant results were obtained. The amplifiers are designed with high precision and their frequency parameters are mainly deviated by parasitic capacitances, the effect of which is reduced by the negative capacitance circuit [61, 121].

In the proposed method, the offset error of the amplifiers undergoing calibration is corrected by feedback (Fig. 4.63).

Correction of offset error is performed in two stages by means of switches. During the first stage, g1 and g2 are connected to the capacitor, and g3 disconnects the input from the output. The input offset (4.17) is established on the capacitor.

4.3 Methods of Improving the Linearity of Signal's Analog-to-Digital...

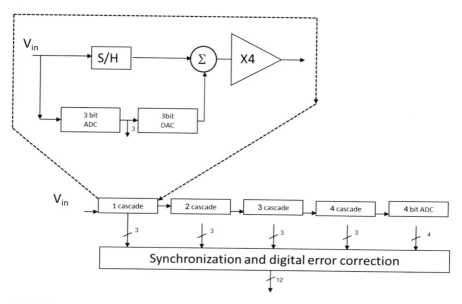

Fig. 4.64 Application of offset correction method in pipeline ADC

$$V_c = \left(\frac{A}{A+1} V_{\text{offset}}\right) \qquad (4.17)$$

During the second stage, g3 is connected to the capacitor, and g1 and g2 are disconnected from the capacitor. Thus, when a signal is applied to the input of the amplifier, V_c voltage on the capacitor is added to it, which is equal to the input offset. Therefore, the offset error is eliminated and the linearity of the amplifier is increased. The method is applicable to all cascade amplifiers (Fig. 4.64).

As a result of applying the method, the linearity of pipeline ADC will increase. After applying the method, simulations were performed. Measurements of INL and DNL were made before (Figs. 4.65 and 4.66) and after (Figs. 4.67 and 4.68) application of the method. The system reduces the nonlinearity of the ADC by approximately 2.5 times (Table 4.6).

Thus, a means of improving the linearity of pipeline ADC with self-calibration has been developed. By reducing the system, the offset error of the amplifiers increases the linearity of the overall system. The implementation of the tool was carried out with the introduction of as few auxiliary elements as possible, and compared to the existing solutions, it has greater efficiency, and less area is required for the implementation of the method. This method is also universal and can be used in pipeline ADCs with different structures with minimal modifications. The system effectively reduces the nonlinearity of the ADC by about 2.5 times. Compared to the previous method, it does not reduce the gain error.

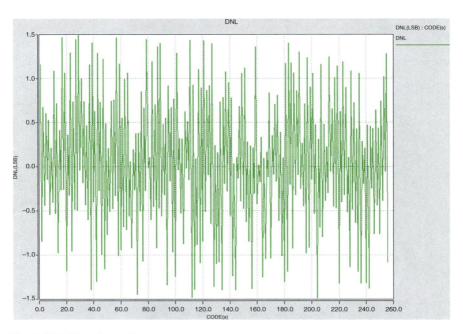

Fig. 4.65 DNL before calibration

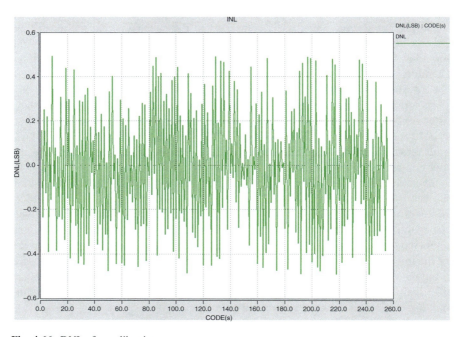

Fig. 4.66 DNL after calibration

4.3 Methods of Improving the Linearity of Signal's Analog-to-Digital... 213

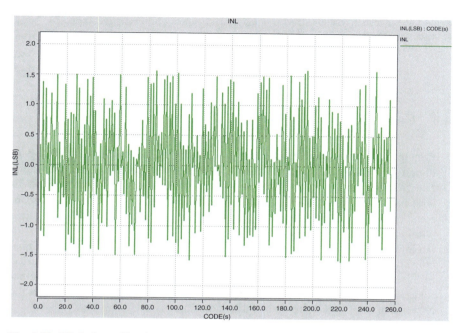

Fig. 4.67 INL before calibration

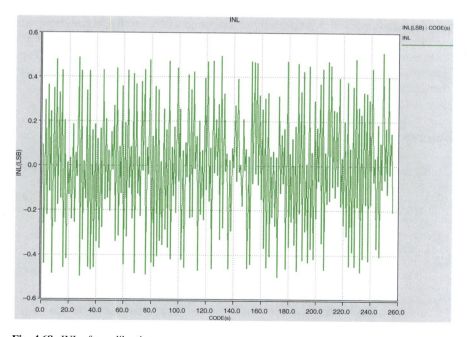

Fig. 4.68 INL after calibration

Table 4.6 Simulation results of the designed pipeline ADC INL (LSB)

Temperature/process	Typical	Slow	Fast
-40	0.41	0.5	0.39
25	0.4	0.45	0.42
125	0.48	0.52	0.47

4.4 Conclusion

1. Principles of development of means for improving the linearity of analog-to-digital conversion with self-calibration have been proposed, which, unlike the known solutions, are more effective in terms of reducing nonlinearity. Systems with built-in feedback allow to ensure a sufficient reduction of deviated parameters in case of an increase in the area occupied on the die and calibration time within the permissible limits.
2. An integrated system for improvement of the linearity of a flash analog-to-digital converter with self-calibration has been proposed, in which, due to the inclusion of a feedback preamplifier and auxiliary digital-to-analog converters for calibration, the nonlinearity parameters were reduced approximately three times, at the expense of only 12% increase in calibration time.
3. An integrated system for improving the linearity of a digital-to-analog converter with self-calibration has been proposed, in which, due to the comparison and calibration of currents of power sources, the nonlinearity parameters were reduced by 20–25%, in the case of an increase of only 5% of the occupied area.
4. A means of improving the linearity of a pipeline analog-to-digital converter with self-calibration has been proposed, in which, due to the implementation of the amplifier offset error reduction system, the nonlinearity parameters were reduced approximately 2.5 times. Compared to the previously existing method, the gain error does not decrease.

References

1. H. Zhou, X. Gui, P. Gao, Design of a 12-bit 0.83 MS/s SAR ADC for an IPMI SoC. 28th IEEE international system-on-chip conference (SOCC) (2015), pp. 175–179
2. S. Saisundar, J.H. Cheong, M. Je, A 1.8 V 1MS/s rail-to-rail 10-bit SAR ADC in 0.18 μm CMOS. 2012 IEEE international symposium on radio-frequency integration technology (RFIT) (2012), pp. 83–85
3. G. Park, M. Song, A CMOS current-steering D/A converter with full-swing output voltage and a quaternary driver. IEEE Trans. Circuits Syst. II Express Briefs **62**, 441–445 (2014)
4. Y. Lan, J. Zhu, C. Wang, A method for compensating the D/A converter frequency response distortion in different nyquist zones. IEEE 2nd international conference on electronics technology (ICET) (2019), pp. 84–87
5. G.I. Radulov, P.J. Quinn, H. Hegt, A. van Roermund, An on-chip self-calibration method for current mismatch in D/A converters. Proceedings of the 31st European solid-state circuits conference (ESSCIRC) (2005), pp. 169–172

References

6. G.A.M. Van Der Plas, A 14-bit intrinsic accuracy Q2 random walk m 90 CMOS DAC. IEEE J. Solid-State Circuits **34**, 12 (1999)
7. Y. Luo, L. Qi, A. Jain, M. Ortmanns, A high-resolution delta-sigma D/A converter architecture with high tolerance to DAC mismatch. IEEE international symposium on circuits and systems (ISCAS) (2018), pp. 1–5
8. S. Kulis, D. Yang, D. Ghong, et al., 26th A high-resolution, wide-range, radiation-hard clock phase-shifter in a 65 nm CMOS technology. International conference "mixed design of integrated circuits and systems" (MIXDES) (2019), pp. 147–150
9. W. Kester, *Data Conversion Handbook* (Engineeri A. D. I., 2005), pp. 976
10. D.K. Jung, Y.H. Jung, T. Yoo, et al.. A 12-bit multi-channel RR DAC using a shared resistor string scheme for area-efficient display source driver. IEEE transactions on circuits and systems I: Regular papers (2018), pp. 3688–3697
11. V. Kommangunta, K. Shehzad, D. Verma, et al., Low-power area-efficient 8-bit coarse-fine resistor-string DAC. IEEE international conference on consumer electronics-Asia (ICCE-Asia) (2020), pp. 1–3
12. B.D. Yang, Y.K. Shin, K.C. Ryu, et al., An area-efficient coarse-fine resistor-string D/A converter. First IEEE Latin American symposium on circuits and systems (LASCAS) (2010), pp. 29–32
13. D. Chen, A 16b 5MSPS two-stage pipeline ADC with self-calibrated technology. International conference on information and computer technologies (ICICT) (2018), pp. 155–158
14. C.W. Lu, P.Y. Yin, M.Y. Lin, A 10-bit two-stage R-DAC with isolating source followers for TFT-LCD and AMOLED column-driver ICs. IEEE Trans. Very Large Scale Integration (VLSI) Syst. **27**, 326–336 (2018)
15. S. Mahdavi, R. Ebrahimi, A. Daneshdoust, et al., A 12 bit 800MS/s and 1.37 mW digital to analog converter (DAC) based on Novel RC technique. IEEE international conference on power, control, signals and instrumentation engineering (ICPCSI) (2017), pp. 163–166
16. J.S. Na, S.K. Hong, O.K. Kwon, A highly linear 10-bit DAC of data driver IC using source degeneration load for active matrix flat-panel displays. IEEE Trans. Circuits Syst. II Express Briefs **67**, 2312–2316 (2020)
17. A. Atanesyan, M. Grigoryan, H. Margaryan, et al., Method of increasing current DAC linearity with considering its random variables for modeling risk or uncertainty. Вестник РАУ **2**, 64–70 (2020)
18. D.A. Johns, K. Martin, *Analog Integrated Circuit Design* (Wiley, New York, 2008), p. 696
19. A. Fayed, M. Ismail, *Adaptive Techniques for Mixed Signal System on Chip* (Springer Science & Business Media, 2006), p. 178
20. T. Shirakawa, R. Sakai, S. Nakatake, On-chip Impedance evaluation with auto-calibration based on auto-balancing bridge. IEEE 61st international midwest symposium on circuits and systems (MWSCAS) (2018), pp. 262–265
21. B. Razavi, The flash adc [a circuit for all seasons]. IEEE Solid-State Circuits Magazine **9**, 9–13 (2017)
22. A. Payra, P. Dutta, A. Sarkar, S.K. Sen, et al., Design of a self regulated flash type ADC with high resolution. Michael Faraday IET International Summit (2015), pp. 591–595
23. R.M. Shende, P.R. Gumble, VLSI design of low power high speed 4 bit resolution pipeline ADC in submicron CMOS technology. Int. J. VLSI Design Commun. Syst. **2**, 81 (2011)
24. D.C. Daly, A.P. Chandrakasan, A 6-bit, 0.2 V to 0.9 V highly digital flash ADC with comparator redundancy. IEEE J. Solid State Circuits **44**, 3030–3038 (2009)
25. P. Ritter, S. Le Tual, B. Allard, et al., Design considerations for a 6 bit 20 GS/s SiGe BiCMOS flash ADC without track-and-hold. IEEE J. Solid State Circuits **49**, 1886–1894 (2014)
26. R.J. Van de Plassche, *CMOS Integrated Analog-to-Digital and Digital-to-Analog Converters* (Springer Science & Business Media, 2013), p. 588
27. E. Sall, M. Vesterbacka, K.O. Andersson, A study of digital decoders in flash analog-to-digital converters. IEEE international symposium on circuits and systems (IEEE Cat. No. 04CH37512) 2004, pp. I–I

28. F. Akopyan, R. Manohar, A.B. Apsel, A level-crossing flash asynchronous analog-to-digital converter. 12th IEEE international symposium on asynchronous circuits and systems (ASYNC'06) (2006), pp. 11–22
29. F. Maloberti, *Analog Design for CMOS VLSI Systems* (Springer Science & Business Media, 2006), p. 374
30. L. Kouhalvandi, S. Aygün, G.G. Özdemiret, et al., 10-bit High-speed CMOS comparator with offset cancellation technique. 5th IEEE workshop on advances in information, electronic and electrical engineering (AIEEE) (2017), pp. 1–4
31. A. Rezapour, H. Shamsi, H. Abbasizadeh, et al., Low power high speed dynamic comparator. IEEE international symposium on circuits and systems (ISCAS) (2018), pp. 1–5
32. S.K. Vinodiya, R.S. Gamad, Analysis and design of low power, high speed comparators in 180 nm technology with low supply voltages for ADCs. 8th international conference on computing, communication and networking technologies (ICCCNT) (2017), pp. 1–5
33. R. Vanitha, S. Thenmozhi, Low power CMOS comparator using bipolar CMOS technology for signal processing applications. 2nd international conference on electronics and communication systems (ICECS) (2015), pp. 1241–1243
34. A. Gupta, A. Agarwal, An efficient fully differential voltage comparator. J. Eng. Sci. Technol., 3162–3172 (2018)
35. S. Velagaleti, A novel high speed dynamic comparator with low power dissipation and low offset (2009), pp. 2–15
36. B.B.A. Fouzy, M.B.I. Reaz, M.A.S. Bhuiyan, et al., Design of a low-power high-speed comparator in 0.13 μm CMOS. International conference on advances in electrical, electronic and systems engineering (ICAEES) (2016), pp. 289–292
37. P. Suriyavejwongs, E. Leelarasmee, W. Pora, A low voltage CMOS current comparator with offset compensation. IEEE Asia Pacific conference on circuits and systems (APCCAS) (2019), pp. 161–164
38. A. Abidi, H. Xu, Understanding the regenerative comparator circuit. Proceedings of the IEEE custom integrated circuits conference (2014), pp. 1–8
39. X. Xin, J. Cai, R. Xie, P. Wang, Ultra-low power comparator with dynamic offset cancellation for SAR ADC. Electron. Lett. **53**, 1572–1574 (2017)
40. J. Zhang, X. Ren, S. Liu, et al., An 11-bit 100-MS/s pipelined-SAR ADC reusing PVT-stabilized dynamic comparator in 65-nm CMOS. IEEE Trans. Circuits Syst. II Express Briefs **67**, 1174–1178 (2019)
41. S. Hanfoug, N.E. Bouguechal, S. Barra, Behavioral non-ideal model of 8-bit current-mode successive approximation registers ADC by using Simulink. Int. J. u-and e-Service, Sci. Technol. **8**, 85–102 (2014)
42. Y.Z. Lin, C.C. Liu, G.Y. Huang, et al., A 9-bit 150-MS/s subrange ADC based on SAR architecture in 90-nm CMOS. IEEE Trans. Circuits Syst. I: Regular Papers **60**, 570–581 (2013)
43. D.H. Lee, T.H. Kuo, K.L. Wen, Low-cost 14-bit current-steering DAC with a randomized thermometer-coding method. IEEE Trans. Circuits Syst. II Express Briefs **56**, 137–141 (2009)
44. E. Greenwald, C. Maier, Q. Wang, et al., A CMOS current steering neurostimulation array with integrated DAC calibration and charge balancing. IEEE Trans. Biomed. Circuits Syst. **11**, 324–335 (2017)
45. A. Narayanan, M. Bengtsson, R. Ragavan, Q.T. Duong, A 0.35 μm CMOS 6-bit current steering DAC. European conference on circuit theory and design (ECCTD) (2013), pp. 1–4
46. T. Chen, G.G.E. Gielen, A 14-bit 200-MHz current-steering DAC with switching-sequence post-adjustment calibration. IEEE J. Solid State Circuits **42**, 2386–2394 (2007)
47. L. Li, M. Xu, X. Huang, et al., A 12 Bit 500MS/s SHA-less ADC in 0.18 um CMOS. IEEE international nanoelectronics conference (INEC) (2016), pp. 1–2
48. H. Zhang, Y. Zhu, C.H. Chan, R.P. Martins, 27.6 A 25MHz-BW 75dB-SNDR inherent gain error tolerance noise-shaping SAR-assisted pipeline ADC with background offset calibration. IEEE international solid-state circuits conference (ISSCC) (2021), pp. 380–382

References

49. J. Oliveira, J. Goes, M. Figueiredo, et al., An 8-bit 120-MS/s interleaved CMOS pipeline ADC based on MOS parametric amplification. IEEE Trans. Circuits Syst. II Express Briefs, 105–109 (2010)
50. S. Devarajan, L. Singer, D. Kelly, et al., A 12-b 10-GS/s interleaved pipeline ADC in 28-nm CMOS technology. IEEE J. Solid State Circuits **52**, 3204–3218 (2017)
51. H. Van de Vel, B.A. Buter, H. van der Ploeg, et al., A 1.2-V 250-mW 14-b 100-MS/s digitally calibrated pipeline ADC in 90-nm CMOS. IEEE J. Solid State Circuits **44**, 1047–1056 (2009)
52. J. Wu, A. Chou, C. H. Yang, et al., A 5.4 gs/s 12b 500 mW pipeline adc in 28nm cmos. Symposium on VLSI circuits (2013), pp. 92–93
53. D. Vecchi, J. Mulder, F. M. van der Goes, et al., An 800MS/s dual-residue pipeline ADC in 40nm CMOS. IEEE international solid-state circuits conference (2011), pp. 184–186
54. C.Y. Chen, J. Wu, J.J. Hung, et al., A 12-bit 3 GS/s pipeline ADC with 0.4 mm2 and 500 mW in 40 nm digital CMOS. IEEE J. Solid State Circuits **47**, 1013–1021 (2012)
55. Understanding Pipelined ADCs. https://pdfserv.maximintegrated.com/en/an/AN1023.pdf
56. T. Liechti, A. Tajalli, O.C. Akgun, et al., A 1.8 V 12-bit 230-MS/s pipeline ADC in 0.18 μm CMOS technology. IEEE Asia Pacific conference on circuits and systems (APCCAS) (2008), pp. 21–24
57. B. Verbruggen, J. Craninckx, M. Kuijk, et al., A 2.6 mW 6 bit 2.2 GS/s fully dynamic pipeline ADC in 40 nm digital CMOS. IEEE J. Solid State Circuits **45**, 2080–2090 (2010)
58. C. Wang, X. Wang, Y. Ding, et al., A 14-bit 250MS/s low-power pipeline ADC with aperture error eliminating technique. IEEE international symposium on circuits and systems (ISCAS) (2018), pp. 1–5
59. D. Miyazaki, M. Furuta, S. Kawahito, A 75 mW 10 bit 120MSample/s parallel pipeline ADC. 29th European solid-state circuits conference (ESSCIRC) (IEEE Cat. No. 03EX705) (2003), pp. 719–722
60. Y.M. Lin, B. Kim, P.R. Gray, A 13-b 2.5-MHz self-calibrated pipelined A/D converter in 3-mu m CMOS. IEEE J. Solid State Circuits **26**, 628–636 (1991)
61. M.T. Grigoryan, A.A. Atanesyan, G.H. Hakobyan, et al., Two stage CTLE for high speed data receiving. IEEE 40th international conference on electronics and nanotechnology (ELNANO) (2020), pp. 374–377
62. C.K. Hsu, T.R. Andeen, N. Sun, A pipeline SAR ADC with second-order interstage gain error shaping. IEEE J. Solid State Circuits **55**, 1032–1042 (2020)
63. J. Li, U.K. Moon, A 1.8-V 67-mW 10-bit 100-MS/s pipelined ADC using time-shifted CDS technique. IEEE J. Solid State Circuits **39**, 1468–1476 (2004)
64. C.K. Hsu, N. Sun, A 75.8 dB-SNDR Pipeline SAR ADC with 2 nd-order interstage gain error shaping. Symposium on VLSI circuits 2019, pp. 68–69
65. J. Zhong, Y. Zhu, S.W. Sin, et al., Inter-stage gain error self-calibration of a 31.5 fJ 10b 470MS/s pipelined-SAR ADC. IEEE Asian solid state circuits conference (A-SSCC) (2012), pp. 153–156
66. H. Yu, M.C.F. Chang, A 1-V 1.25-GS/S 8-bit self-calibrated flash ADC in 90-nm digital CMOS. IEEE Trans. Circuits Syst. II Express Briefs **55**, 668–672 (2008)
67. J. Sun, J. Wu, A self-calibrated multiphase timing system in time-interleaved ADC. IEEE 2nd advanced information technology, electronic and automation control conference (IAEAC) (2017), pp. 292–295
68. M.S. Reddy, S.T. Rahaman, An effective 6-bit flash ADC using low power CMOS technology. 15th international conference on advanced computing technologies (ICACT) (2013), pp. 1–4
69. A.S.T.H.F. Kuttner, C. Sandner, M. Clara, A 6bit, 1.2 gsps low-power flash-adc in 0.13 μm digital cmos. IEEE J. Solid State Circuits, 111–115 (2005)
70. S. Park, Y. Palaskas, A. Ravi, et al., A 3.5 GS/s 5-b flash ADCin 90 nm CMOS. IEEE custom integrated circuits conference (2006), pp. 489–492

71. P.V. Rahul, A.A. Kulkarni, S. Sankanur, et al., Reduced comparators for low power flash adc using tsmc018. International conference on microelectronic devices, circuits and systems (ICMDCS) (2017), pp. 1–5
72. T. M. Ignatius, J. K. Antony, S. R. Mary, Implementation of high performance dynamic flash ADC. Annual international conference on emerging research areas: magnetics, machines and drives (AICERA/iCMMD, IEEE) (2014), pp. 1–5
73. H. Fan, J. Li, Q. Feng, et al., Exploiting smallest error to calibrate non-linearity in SAR ADCs. IEEE Access (2018), pp. 42930–42940
74. D.S. Shylu, S. Radha, P.S. Paul, P.S. Sudeepa, Design of low power 4-bit Flash ADC in 90nm CMOS process. 2nd international conference on signal processing and communication (ICSPC) (2019), pp. 252–257
75. A. Van Roermund, M. Vertregt, D. Leenaerts, et al., A 12 b 500 MS/s DAC with> 70dB SFDR up to 120 MHz in 0.18 μm CMOS. IEEE international digest of technical papers. solid-state circuits conference (ISSCC) (2005), pp. 116–117
76. G. Radulov, P. Quinn, A 0.037 mm2 1GSps 12b self-calibrated 40nm CMOS DAC cell with SFDR> 60 dB up to 200 MHz and IM3<—60dB up to 350MHz. European conference on circuit theory and design (ECCTD) (2020), pp. 1–4
77. A.R. Bugeja, B.S. Song, A self-trimming 14-b 100-MS/s CMOS DAC. IEEE J. Solid State Circuits **35**, 1841–1852 (2000)
78. D. Arbet, G. Nagy, V. Stopjaková, G. Gyepes, A self-calibrated binary weighted DAC in 90nm CMOS technology. 29th international conference on microelectronics proceedings (MIEL) (2014), pp. 383–386
79. J.H. Chi, S.H. Chu, T.H. Tsai, A 1.8-V 12-bit 250-MS/s 25-mW self-calibrated DAC. Proceedings of ESSCIRC (2010), pp. 222–225
80. T. Rabuske, J. Fernandes, F. Rabuske, et al., A self-calibrated 10-bit 1 MSps SAR ADC with reduced-voltage charge-sharing DAC. IEEE international symposium on circuits and systems (ISCAS) (2013), pp. 2452–2455
81. H.W. Chen, W.T. Shen, W.C. Cheng, H.S. Chen, A 10b 320MS/s self-calibrated pipeline ADC. IEEE Asian solid-state circuits conference (2010), pp. 1–4
82. S.K. Gupta, M.A. Inerfield, J. Wang, A 1-GS/s 11-bit ADC with 55-dB SNDR, 250-mW power realized by a high bandwidth scalable time-interleaved architecture. IEEE J. Solid State Circuits **41**, 2650–2657 (2006)
83. Y. Chen, J. Wang, H. Hu, et al., A 200 MS/s, 11 bit SAR-assisted pipeline ADC with bias-enhanced ring amplifier. IEEE international symposium on circuits and systems (ISCAS) (2017), pp. 1–4
84. Y.D. Jeon, Y.K. Cho, J.W. Nam, et al., A 9.15 mW 0.22 mm 2 10b 204MS/s pipelined SAR ADC in 65nm CMOS. IEEE custom integrated circuits conference (2010), pp. 1–4
85. W. Li, F. Li, C. Yang, et al., An 85 mW 14-bit 150 MS/s pipelined ADC with a merged first and second MDAC. China communications (2015), pp. 14–21
86. S.M. Louwsma, A.J.M. van Tuijl, M. Vertregt, R. Nauta, A 1.35 GS/s, 10 b, 175 mW time-interleaved AD converter in 0.13 μm CMOS. IEEE J. Solid State Circuits **43**, 778–786 (2008)
87. K. Gulati, M.S. Peng, A. Pulincherry, et al., A highly integrated CMOS analog baseband transceiver with 180 MSPS 13-bit pipelined CMOS ADC and dual 12-bit DACs. IEEE J. Solid State Circuits **41**, 1856–1866 (2006)
88. A. Verma, B. Razavi, A 10-bit 500-ms/s 55-mW cmos adc. IEEE J. Solid State Circuits **44**, 3039–3050 (2009)
89. S.H.W. Chiang, H. Sun, B. Razavi, A 10-bit 800-mhz 19-mw CMOS adc. IEEE J. Solid State Circuits **49**, 935–949 (2014)
90. B.D. Sahoo, B. Razavi, A 10-b 1-GHz 33-mW CMOS ADC. IEEE J. Solid State Circuits **48**, 1442–1452 (2013)
91. A.M. Ali, H. Dinc, P. Bhoraskar, et al., A 14-bit 2.5 GS/s and 5GS/s RF sampling ADC with background calibration and dither. IEEE symposium on VLSI circuits (VLSI-Circuits) (2016), pp. 1–2

References

92. C.H. Chan, Y. Zhu, Z. Zheng, et al., A 39mW 7b 8GS/s 8-way TI ADC with cross-linearized input and bootstrapped sampling buffer front-end. 44th European solid state circuits conference (ESSCIRC) (2018), pp. 254–257
93. Y. Haque, D.E. Lewis, R. Hales, et al., Time interleaved 16 bit, 250MS/s ADC using a hybrid voltage/current mode architecture with foreground calibration. 40th European solid state circuits conference (ESSCIRC) (2014), pp. 59–62
94. C.H. Chan, Y. Zhu, W.H. Zhang, et al., A two-way interleaved 7-b 2.4-GS/s 1-then-2 b/cycle SAR ADC with background offset calibration. IEEE J. Solid State Circuits **53**, 850–860 (2018)
95. C.H. Chan, Y. Zhu, S.W. Sin, et al., 26.5 A 5.5 mW 6b 5GS/S 4×−Interleaved 3b/cycle SAR ADC in 65nm CMOS. IEEE international solid-state circuits conference (ISSCC) Digest of Technical Papers (2015), pp. 1–3
96. M. Baert, W. Dehaene, 20.1 a 5GS/s 7.2 ENOB time-interleaved VCO-based ADC achieving 30.5 fJ/conv-step. IEEE international solid-state circuits conference (ISSCC) (2019), pp. 328–330
97. A. Ramkaj, J.C.P. Ramos, Y. Lyu, et al., 3.3 A 5GS/s 158.6 mW 12b Passive-Sampling 8×-Interleaved Hybrid ADC with 9.4 ENOB and 160.5 dB FoM S in 28nm CMOS. IEEE international solid-state circuits conference (ISSCC) (2019), pp. 62–64
98. B. Vaz, B. Verbruggen, C. Erdmann, et al., A 13bit 5GS/s ADC with time-interleaved chopping calibration in 16nm FinFET. 2018 IEEE symposium on VLSI circuits (2018), pp. 99–100
99. M.B. Dayanik, D. Weyer, M.P. Flynn A 5GS/s 156MHz BW 70dB DR continuous-time sigma-delta modulator with time-interleaved reference data-weighted averaging. 2017 symposium on VLSI circuits (2017), pp. 38–39
100. Z. Yu, D. Chen, Algorithm for dramatically improved efficiency in ADC linearity test. IEEE international test conference (2012), pp. 1–10
101. B. Chen, M. Maddox, M.C. Coln, et al., Precision passive-charge-sharing SAR ADC: Analysis, design, and measurement results. IEEE J. Solid State Circuits **53**, 1481–1492 (2018)
102. L. Jin, K. Parthasarathy, T. Kuyel, et al., Accurate testing of analog-to-digital converters using low linearity signals with stimulus error identification and removal. IEEE Trans. Instrum. Meas. **54**, 1188–1199 (2005)
103. L. Jin, D. Chen, R.L. Geiger, SEIR linearity testing of precision a/D converters in nonstationary environments with center-symmetric interleaving. IEEE Trans. Instrum. Meas. **56**, 1776–1785 (2007)
104. S. Kook, H.W. Choi, A. Chatterjee, Low-resolution DAC-driven linearity testing of higher resolution ADCs using polynomial fitting measurements. IEEE transactions on very large scale integration (VLSI) systems (2012), pp. 454–464
105. H. Xu, L. Wang, R. Yuan, Y. Chang, A/D converter background calibration algorithm based on neural network. International conference on electronics technology (ICET IEEE) (2018), pp. 1–4
106. H. Xu, L. Wang, R. Yuan, Y. Chang, Combined spectral and histogram analysis for fast ADC testing. IEEE Trans. Instrum. Meas. **54**, 1617–1623 (2005)
107. H.M. Chang, C.H. Chen, K.Y. Lin, K.T. Cheng, Calibration and testing time reduction techniques for a digitally-calibrated pipelined ADC. 27th IEEE VLSI test symposium (2009), pp. 291–296
108. J. Duan, L. Jin, D. Chen, INL based dynamic performance estimation for ADC BIST. Proceedings of 2010 ieee international symposium on circuits and systems (2010), pp. 3028–3031
109. T. Liu, L. Chen, L. Liu, et al., A calibration method of SFDR based on INL for pipelined A/D converters. IEEE international conference on electron devices and solid-state circuits (2014), pp. 1–2

110. D.R. Oh, J.I. Kim, D.S. Jo, et al., A 65-nm CMOS 6-bit 2.5-GS/s 7.5-mW 8$\times$ time-domain interpolating flash ADC with sequential slope-matching offset calibration. IEEE J. Solid State Circuits, 288–297 (2018)
111. H. Tang, H. Zhao, S. Fan, et al., Design technique for interpolated flash ADC. 10th IEEE international conference on solid-state and integrated circuit technology (2010), pp. 180–183
112. H.C. Lee, J.A. Abraham, A novel low power 11-bit hybrid ADC using flash and delay line architectures. Design, automation & test in Europe conference & exhibition (DATE IEEE) (2014), pp. 1–4
113. J. Wu, F. Li, W. Li et al., A 14-bit 200MS/s low-power pipelined flash-SAR ADC. IEEE 58th international midwest symposium on circuits and systems (MWSCAS) (2015), pp. 1–4
114. Y.K. Cho, J.H. Jung, K.C. Lee, A 9-bit 100-MS/s flash-SAR ADC without track-and-hold circuits. International symposium on wireless communication systems (ISWCS) (2012), pp. 880–884
115. P. Dhage, P. Jadhav, Design of power efficient hybrid flash-successive approximation register analog to digital converter. International conference on communication and signal processing (ICCSP) (2017), pp. 462–466
116. S. Fan, H. Zhao, H. Tang, et al., Mixed AC/DC-coupled averaging technique for ADC nonlinearity reduction. 2nd Asia symposium on quality electronic design (ASQED) (2010), pp. 102–105
117. J. Ren, J. Xiong, J. Liu, High-speed ADC quantization with overlapping metastability zones. IEEE 61st international midwest symposium on circuits and systems (MWSCAS) (2018), pp. 234–237
118. M. Miyahara, I. Mano, M. Nakayama, et al., 22.6 A 2.2 GS/s 7b 27.4 mW time-based folding-flash ADC with resistively averaged voltage-to-time amplifiers. IEEE international solid-state circuits conference digest of technical papers (ISSCC) (2014), pp. 388–389
119. A.A. Atanesyan, An on-chip self-calibration method for 8-bit flash ADC. Proceedings of the Republic of Armenia National Academy of Sciences and National Polytechnic University of Armenia: Series of technical sciences (2021), pp. 75–82
120. V.S. Melikyan, A.A. Atanesyan, M.T. Grigoryan et al., Duty-cycle correction circuit for high speed interfaces. IEEE 39th international conference on electronics and nanotechnology (ELNANO) (2019), pp. 42–45
121. A. Atanesyan, CMOS negative capacitance with improved AC performance. RAU manual (2019), pp. 49–58

Chapter 5
Design of High-performance Heterogeneous Integrated Circuits

5.1 General Issues of Designing Means for High-performance Heterogeneous Integrated Circuits

5.1.1 Importance of Design Means for High-performance Heterogeneous Integrated Circuits

Currently, in the period of development of high-performance computing systems, the number of multi-core integrated circuits (ICs), which is vital in these systems, is constantly increasing [1]. The number of transistors used in ICs and logarithmic dependence in different periods (Fig. 5.1) [2] increases by a nonlinear law.

The development of ICs is progressing according to Moore's law [3]. With each change in the technological process, the number of transistors in ICs is doubled, and the power consumption is maintained on the account of reducing the supply voltage. However, it is not possible to constantly reduce it in order to limit the increase of the leakage current of transistors [4]. And because of these problems, it became necessary to introduce the possibility of parallelism of operations performed through IC in order to ensure the growth of computing performance [5–8].

On the other hand, the performance of processor systems directly depends on temperature and power consumption [9]. In order to limit the dissipating thermal energy that can be turned off, as well as the temperature increase as a result, energy consumption control mechanisms are used, the principle of which is the limitation of the maximum frequency of IC [10]. The latter causes a drop in performance [11]. To overcome this limitation, multi-core central processor units (CPUs) were created, which made it possible to have more than one computing core in one IC [12]. Over the years, since the number of IC cores, power consumption, and frequency changes (Fig. 5.2) [2], when the maximum frequency value in ICs approached 4–5 GHz, and the power consumption reached 100–150 W, multi-cores began to be designed.

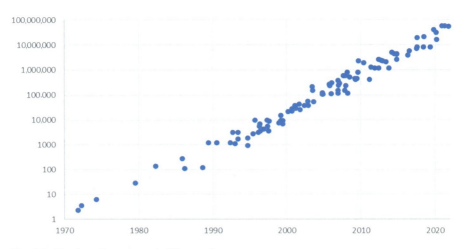

Fig. 5.1 Number of transistors in ICs over the years

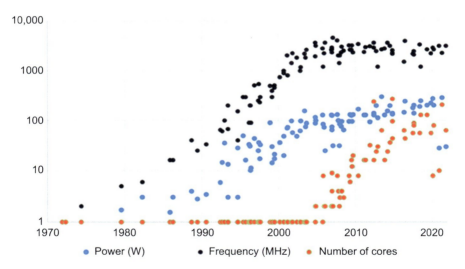

Fig. 5.2 The change of IC parameters over the years

Single-core CPUs are not scalable from an architectural perspective. It is not possible to increase performance along with power consumption. However, multi-core architectures can be expanded up simply by increasing the number of cores. This is also confirmed by Polak's principle [13], according to which the performance of the core changes according to the square law depending on the complexity of the implementation. Along with the complexity of the implementation, the power consumption and core area increase. Multi-core system performance increase (MSPI) is limited by the percentage of the sequential program in the total one and is determined by the following formula [14]:

$$\text{MSPI} = \frac{1}{(\text{seq}\% + (1 - \text{seq}\%)/N)} \tag{5.1}$$

where N is the number of cores. The higher the percentage of the sequential program in the total one, the lower the MSPI saturates, depending on the number of cores.

As mentioned above, multi-core CPUs also provide parallelism at the instruction execution level. For example, a processor consisting of two cores in the case of the same frequency can provide up to twice higher performance compared to a single-core processor [15, 16]. Currently, single-core processors are used only in systems where there is no need for high computing power.

The architectures of multi-core processors in turn have certain limitations. Interconnection logic circuits in processors limit the bandwidth of data transfer between cores. The larger the number of cores, the larger interconnection circuit is needed to ensure the connection between them [17, 18]. For example, in order to provide more than eight cores in one IC, one needs to have a network in IC instead of interconnection [19].

There are two types of multi-core ICs. The first are processors that have several cores and a first- or second-level shared cache memory [20]. The above examples belong to this class. The second is those ICs where there are hundreds or thousands of cores that are logically connected to each other through a network [21]. In these systems, groups of several cells can form clusters. Clusters usually consist of one or more cores, and these cores have shared memory. An example of such ICs is graphs processor units (GPUs) [22]. GPUs are used to display an image on a monitor by processing parallel three-dimensional objects. In order to solve this problem, GPUs are designed with small computing cores. These cores are not as compatible as processors, but enable the programmer to perform mathematical calculations and perform them in parallel [23]. Current GPUs include tens of thousands of cores. However, such systems also have certain limitations. Here, all cores are identical and perform the same type of operations.

Along with all this, there is a limit to the number of concurrently operating cores [24]. When all cores are in active mode at the same time, the power consumption of IC increases significantly [25]. Its increase leads to such temperature values, in which the efficiency of IC operation decreases. This problem is solved by limiting the number of concurrently active cores or by reducing the operating frequency. In both cases, IC performance decreases.

One of the existing solutions to these limitations as problems is the application of heterogeneous systems, when computing cores of different types and with instructions or data parallelism are used [26]. A heterogeneous IC can consist of cores with the same instruction set but different microarchitectures, or cores with different instruction sets, or cores-accelerators that do not have an instruction set [27]. And special-purpose cores allow to ensure higher performance compared to compatible cores at the same power cost. Thus, when performing calculations using heterogeneous ICs, it is possible to achieve greater productivity while maintaining power consumption, or in the case of a lower value of the latter, while maintaining performance.

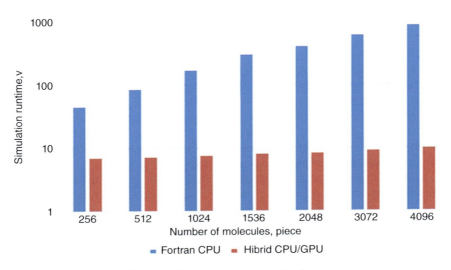

Fig. 5.3 Dependence of simulation time and number of molecules

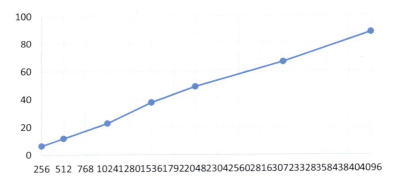

Fig. 5.4 Dependence of comparison of simulation speed of molecules and number

Heterogeneous systems are used in telecommunications, space devices, cryptocurrency computing, and other fields. For example, for hashing calculations of cryptocurrencies, such systems are used, which make it possible to perform the calculations in parallel [28]. In computers, in order to increase the performance of parallel calculations, programmers use computing power of GPUs. The presented graph (Fig. 5.3) shows the increase in performance when applying a heterogeneous system in molecule simulation applications [29].

Simulation was performed on a network of parallel processors (Fortran CPU) and through a heterogeneous system (CPU/GPU). As seen from the dependence of the number of molecules and the simulation time (Fig. 5.4), its time is reduced by 87 times when performing simulation through a heterogeneous system [29].

5.1 General Issues of Designing Means for High-performance...

In summary, the factors that cause the demand for the design of heterogeneous ICs can be mentioned.

Flexible architecture. Heterogeneous ICs are multi-core computing systems; for the implementation of which it is necessary to design special-purpose and compatible cores. Compatible cores will enable complex algorithms to be implemented at the programming level, and special-purpose cores for low-power and high-performance calculations. Optimal resource allocation will ensure maximum performance.

Compatibility and reduction of power consumption. As mentioned above, improving compatibility leads to increase in power consumption. For this reason, heterogeneous ICs are used to relieve the burden of cores in VLSI. The improvement of design problems of the latter and the availability of flexible architecture contribute to the reduction of IC compatibility and the complexity of the implementation of cores, which in turn leads to the reduction of power consumption.

Reduction of design time. The design of special ICs, depending on the field of application, may have different features due to the presence of different computing nodes and interfaces. And with a flexible architectural heterogeneous IC design tool, design time can be significantly reduced by using common interfaces and configurable components.

Structure of High-performance Heterogeneous Integrated Circuits

There are four types of processor systems according to Michael Flynn's classification [30]:

- Single instruction, single data (SISD)
- Single instruction, multiple data (SIMD)
- Multiple instruction, single data (MISD)
- Multiple instruction, multiple data (MIMD)

SISD—systems with this architecture consist of one single-core processor that executes one instruction at a time, and data is read from one memory [31]. The SISD architecture (Fig. 5.5) corresponds to Von Neumann architecture. Such architectures have parameters for executing asynchronous instructions [32]. Pipelined or superscalar processors are implemented with SISD architecture. Instructions are received from the control unit, decoded and executed by the processor, which reads the data, processes it according to the instruction, and records it in memory.

Superscalar processors are CPUs with instruction-level parallelism. Unlike the classic Von Neumann architecture, which can execute one instruction during one period of the clock signal, superscalar processors have the ability to execute more than one instruction in one clock cycle [33]. This is done in the CPU by integrating more than one instruction execution units, the operation of which is independent of each other and in parallel. It is important to note that each instruction execution unit is not a separate processor. The single-core superscalar processor belongs to the SISD class [34].

Fig. 5.5 SISD architecture

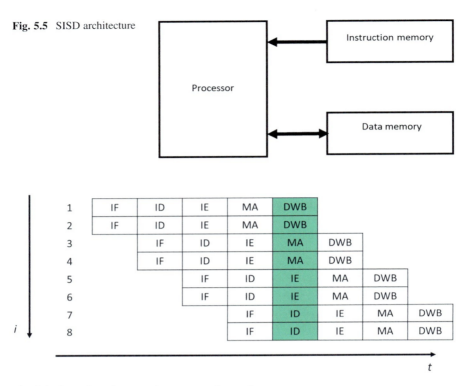

Fig. 5.6 Execution of superscalar processor instructions

As seen from the instruction execution graph of a single-core superscalar processor consisting of two instruction execution units over time (Fig. 5.6), instruction execution is pipelined and consists of several stages. These stages are instruction fetch (IF), instruction decoding (ID), instruction execution (IE), memory access (MA), and data writeback (DWB).

SISD architectures are now being used in microprocessor systems, where power consumption requirements take priority over performance. This architecture is quite simple in terms of implementation, which makes it cheap to manufacture, but uncompetitive in terms of performance compared to other architectures [35].

SIMD—CPUs designed with this architecture execute the same instruction by applying the operation corresponding to the instruction on multiple data. Data is divided into streams here. Instructions are executed in parallel or sequentially. In order to use instructions in parallel, it is necessary to have many processor nodes in a given processor system [36]. Sequential instruction execution is also pipelined, as in the case of SISD. Circuits designed with this architecture are divided into two groups:

5.1 General Issues of Designing Means for High-performance...

- Vector processor (VP).
- Pipelined SIMD processor (PSP).

VPs contain multiple cores, and each core has its own cache memory and register file (Fig. 5.7) [37]. Each VP core is compatible and capable of running different instructions. Instructions are executed in parallel. The number of simultaneous instructions is limited by the number of cores. The structure of one core of VP is presented in Fig. 5.7 [38].

The components of the VP core are algebraic logic unit (ALU), central control unit (CCU), and cache memory and register file (RF). CCU is doing instruction fetch, decoding, and managing data flow. Cache memory is an intermediate unit between RF and system memory (RAM). The structure of VP consisting of four cores is presented in Fig. 5.8 [39].

It can be seen from the diagram that the same instruction is sent to all VP cores (VPC) at the same time, and the processed data is read from the data array in parallel. The execution of the below assembly program over time using VP is presented in Fig. 5.9 [38].

1) LOAD   ARR, IN[3:0]   – read IN array
2) SUB   ARR, 10   – subtract 10 from the read value
3) DIV   ARR, 3   – divide the result by 3
4) STORE   OUT[3:0], ARR   – record the result in the OUT array

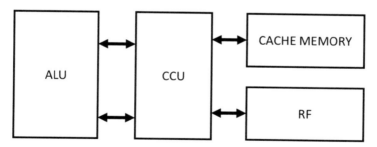

Fig. 5.7 Structure of VP core

Fig. 5.8 Four-stage VP structure

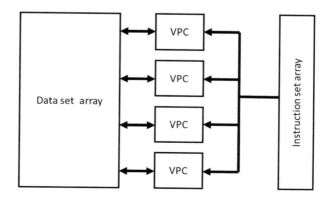

Fig. 5.9 Execution of instructions by quad-core VP

1	LOAD0	LOAD1	LOAD2	LOAD3
2	SUB0	SUB1	SUB2	SUB3
3	DIV0	DIV1	DIV2	DIV3
4	STORE0	STORE1	STORE2	STORE3

Fig. 5.10 Execution of instructions by quad-core pipelined SIMD processor

	LOAD core	SUB core	DIV core	STORE core
1	LOAD0	-	-	-
2	LOAD1	SUB0	-	-
3	LOAD2	SUB1	DIV0	-
4	LOAD3	SUB2	DIV1	STORE0
5	-	SUB3	DIV2	STORE1
6	-	-	DIV3	STORE2
7	-	-	-	STORE3

As it can be seen, different cores execute the same instruction at the same time. The same instruction set is executed in a different way through pipelined SIMD processor. Unlike VP, pipelined SIMD processor cores are not mutually exclusive, and each core executes one unique instruction. Execution of the above-mentioned program by pipelined SIMD processor [38] is presented in Fig. 5.10.

As shown, with pipelined SIMD processor, the instructions are executed sequentially, and unlike VP, the same instructions are executed in the same core, different instructions are executed at the same time.

The advantage of the pipelined SIMD processor architecture compared to VP is that the cores are not compatible and execute unique instructions, due to which they are simpler from a structural point of view, occupy small area, and have low power consumption. To increase the performance of the pipelined SIMD processor, the number of cores executing the same instructions is increased [40].

SIMD architectures are used in GPU ICs, video processing applications, machine learning, as well as heterogeneous ICs [41].

ICs with the MISD architecture execute different instructions on the same data. They are not widely used in consumer electronics and are mainly used in space systems where the probability of errors is high and the same processing is performed by more than one core [42]. The structure of the MISD processor is presented in Fig. 5.11 which is done for the purpose that the failure of one circuit in space devices does not affect the overall system.

Processors with the MIMD structure consist of compatible cores, each of which in turn has a separate instruction array and data array. Due to their structure, these ICs provide maximum performance, but in other parameters they are inferior to all the

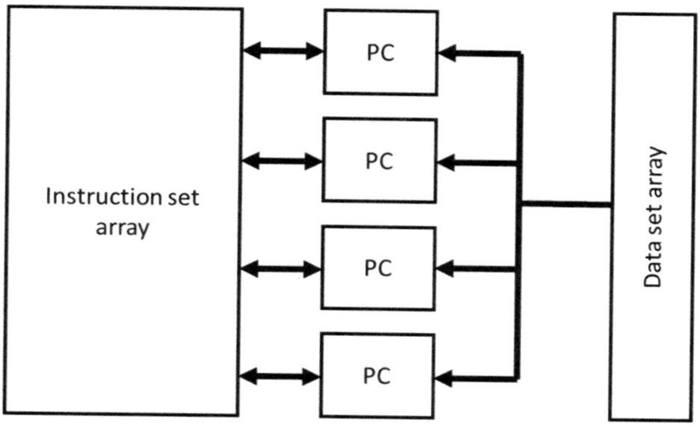

5.1 General Issues of Designing Means for High-performance... 229

Fig. 5.11 Structure of MISD processor

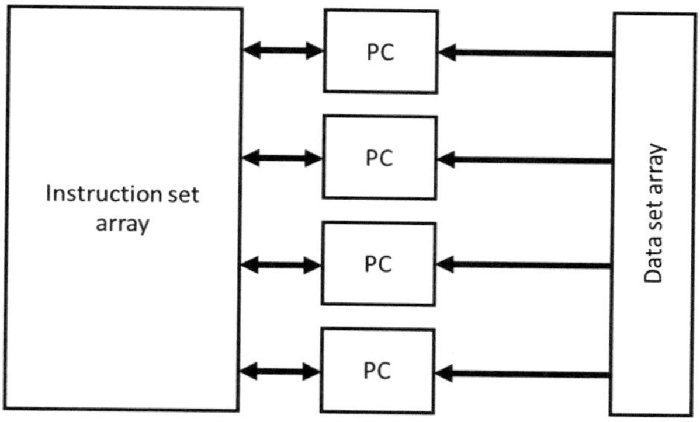

Fig. 5.12 MIMD architecture

above-mentioned architectures [43]. As seen from the MIMD structure (Fig. 5.12), it consists of independent cores that receive independent instructions and perform operations based on an independent data stream or data stream in shared memory [44].

MIMD architectures are used in modern personal computers (PCs), mobile computers, and SoCs of smartphones. They are suitable for multi-flow applications and data-intensive tasks [35].

Design Issues of High-performance Heterogeneous Integrated Circuits

The most important issues for the design of heterogeneous ICs are:

- Implementation of a flexible or configurable architecture
- Data transfer issues between clock domains
- Implementation of data transfer interfaces

The first step in the design of heterogeneous ICs is the choice of architecture. It exclusively depends on the field of application of the given heterogeneous IC, in particular, on the modern computing requirements in that field. For example, network devices are mainly smart systems that require processing of data stream supplied by sensors but have very strict requirements in terms of power consumption. Here, power consumption is preferred over the above requirements. It is desirable to choose such an architecture, which will consist of one or more ASIP cores [45]. In another example, in the case of video processing performed in cloud systems, the advantage is given to performance requirements, because these systems have constant power supply and external cooling devices [46]. In order to obtain maximum performance, it is advised to choose an architecture that provides high parallelism at the data and instruction level. In conclusion, the choice of architecture strongly depends on the field of application, and in order to reduce the design time, there is a need to create a software environment that will enable the generation of a heterogeneous IC with the required architecture from the existing design tools. In those design means, depending on technical requirements, it is necessary to realize the optimal data flow and ensure the interconnections between the components with appropriate interfaces [47].

Another important task is to meet power consumption requirements. For this, it is necessary that the implemented IC has low power consumption modes, when the main computing cores are in an inactive state, and the central control unit, whether automatic or CPU, should have a wide operating frequency range. By lowering the frequency, the supply voltage and therefore the power consumption can be reduced. In addition to low-power operating modes, a heterogeneous IC should have a low-performance mode, when both cores and computing nodes operate at low frequencies, which also leads to a decrease in power consumption [48].

Based on the requirements of power consumption and the characteristics of the structure of heterogeneous ICs, an important design issue is to ensure uninterrupted and as fast as possible data transfer between domains that are asynchronous to each other. Computing cores can have different operating frequencies.

Thus, one of the most relevant problems of modern IC creation is the development of heterogeneous IC system design tools to significantly improve their main parameters and shorten the design period.

5.1.2 Current State and Issues of Design Means for High-performance Heterogeneous Integrated Circuits

The demand for designing high-performance heterogeneous integrated circuits is increasing sharply. This growth is due to the increase in the amount of data in various fields. The number of pixels in camera matrices increases with the development of technological processes, and in the field of video processing, the data of each pixel needs to be processed. Therefore, as the number of pixels increases, so does the amount of data that must be processed, passing that data through various digital logic filters, converting one image representation format to another, and so on. The processing of that data involves performing mathematical operations on the pixel data, such as addition, subtraction, multiplication, division, matrix multiplication, etc. These operations are expensive in terms of power consumption and require parallel processing of video image to be able to meet today's strict requirements [49]. Currently, IC systems of mobile phones are integrating ASICs to perform video processing. As mentioned above, the increase in the amount of data requires the development of tools for designing heterogeneous architectures.

An inseparable part of heterogeneous architectures are control units of data transfer interfaces. These interfaces can be used both for data transfer between internal nodes and between devices externally connected to ICs. One of the widely used interfaces is universal asynchronous receiver-transmitter (UART) [50]. It is designed to transfer data between external devices. UART interface is used in such devices in which data transfer speed does not play a primary role [51]. However, this interface is not flexible in terms of application. The operating frequency is fixed and does not change over time. Basically, devices equipped with UART interface are used with default settings and are not subject to change. Such interfaces are used to exchange data between control registers in heterogeneous ICs (and not only) [52].

Each processor node in computer systems must be connected to the shared system memory through a common switch. The connection between system memory and heterogeneous IC should be optimized without additional applications and efficiently. In modern ICs, the connection between RAM and other constituent nodes is provided by the AMBA AXI interface [53].

In heterogeneous ICs, each node has its own external interfaces, and depending on the operating mode, their frequencies are different, and clock signals are asynchronous [54]. Circuits with asynchronous clock signals form asynchronous clock domains, between which the data transfer must be carried out through synchronizers. Synchronizers are digital circuits that provide lossless data transfer from one clock domain to another but cause additional delay. When using a classical synchronizer, this delay increases with the reduction of technological process and the increase of the frequency in IC which leads to a decrease in data throughput and performance [55].

Data Transfer Problems Between Component Parts in High-performance Heterogeneous ICs

Heterogeneous ICs have many control units, the operating mode of which is based on the values of their control registers. As mentioned above, with the increase in the number of cores in heterogeneous ICs, their structure becomes more complicated, as the logic circuit of the interconnect increases. Part of it is the interfaces providing data transfer between cores and the programming of control registers. Programming of control registers is not a frequent process and is not performed in active operating mode. It is performed once after the circuit is released or before entering or exiting low power mode. However, even in these problems, modern ICs implement parallel interfaces with 32-bit resolution, which occupy a large area on the die. It is recommended to use UART interface for programming the control registers in ICs, through which only one bit is needed for one-way data transfer [56]. It is widely used for data transfer between different devices [57].

A UART is a computing device hub designed to communicate with other digital devices. A node transmits data sequentially.

The parameters that characterize the UART are:

- Full bidirectional data exchange (sequential transmission and reception of data independently of each other)
- Asynchronous data exchanges
- Large performance range
- Provision of 5-, 6-, 7-, and 8-bit long packets
- Provision of 1 or 2 "end" bits
- Provision of even or odd parity bit
- Data loss detection
- Packet structure mismatch error detection
- Protection against false "start" bit

The UART interconnection circuit is presented in Fig. 5.13, where RX is the receiver and TX is the transmitter [58].

In the structure of the presented UART controller, two main constituent parts are divided by dotted lines: transmitter and receiver (Fig. 5.14). And the control registers are common. The receiver consists of a node that detects the negative edge of the voltage on the UART RX line; a clock signal generator, the frequency of which is determined by the value written in control registers; a module that reads 1 bit; a control block; and an RX memory for temporarily storing the received data. The transmitter consists of a TX memory that temporarily stores the data to be sent; a node that generates the clock signal, which is structurally similar to the corresponding node of the receiver; and a control logic node that controls the

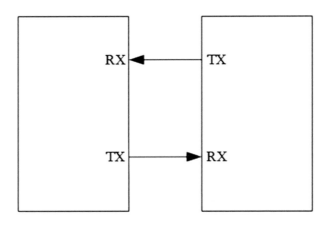

Fig. 5.13 Interconnection of UART controllers

5.1 General Issues of Designing Means for High-performance...

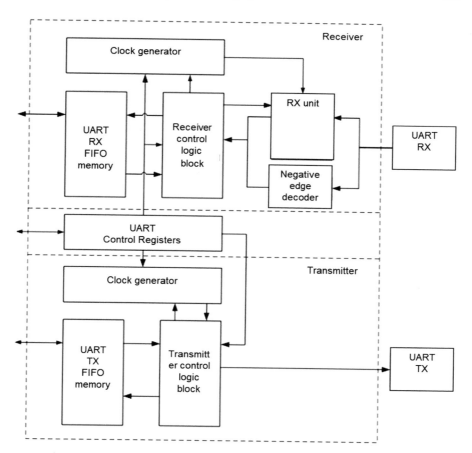

Fig. 5.14 Structure of UART controllers

UART TX line. The transmitter supports the same packet format as the transmitter and can detect packet structure error, data loss error, and parity bit error.

The clock signal generators of the receiver and the transmitter do not differ from each other in structure. The only difference is the different control signals. In other words, these nodes work independently of each other, which ensures full two-way data flow.

The clock signal generator represents a counter, the maximum value of which determines the corresponding output of the control register. Depending on this value, the maximum value of the counter changes, and when it is reached, the counter resets to zero. The operating frequency of the UART is determined by the maximum value of this counter. In terms of frequency selection, the UART controller has 20 operating states.

Table 5.1 UART operating states and performance

Operating mode	Speed (bit/s)	Maximum counter value	Operating mode	Speed (bit/s)	Maximum counter value
0	300	33,333	10	28,800	347
1	600	16,666	11	38,400	260
2	1200	8333	12	57,600	173
3	1800	5555	13	76,800	130
4	2400	4166	14	115,200	86
5	4800	2083	15	128,000	78
6	7200	1388	16	230,400	43
7	9600	1041	17	250,000	40
8	14,400	694	18	500,000	20
9	19,200	520	19	1,000,000	10

Fig. 5.15 UART packet structure

The operating states of the UART controller and performance, corresponding to these states, the maximum value of the clock signal generator counter, corresponding to each state are presented in Table 5.1.

A single transaction packet consists of a single symbol consisting of eight data bits and matching bits ("start" and "end"). The packet may also contain a redundant parity bit. The UART accepts packets with all possible options of parameters listed below as true:

- 1 "start" bit
- Data bit 5, 6, 7, or 8
- Odd or even parity bit or its absence
- 1, 1.5, or 2 "end" bit

The packet begins with the "start" bit, followed by the least significant data bit. After that, the remaining data bits are transmitted sequentially, up to a maximum of eight, ending with the most significant bit. If the parity bit is present in the packet, the data bits are followed by the parity bit, followed by the "end" bit. When a complete packet is sent, it may be immediately followed by a new packet, or a wait state will be established on the line. All possible packet formats are depicted in Fig. 5.15. Bits in square brackets may be present or vice versa [58].

- St—"start" bit, logical "0"
- (n)—data bits (5–8 bits)
- P—parity bit (even or odd)
- Sp—"end" bit, always logical "1"
- IDLE—standby state, always logical "1"

5.1 General Issues of Designing Means for High-performance...

Packet formats are also determined by control registers.

As seen in the packet, there is a parity bit. The parity bit is calculated as follows: XOR logic operation is performed with all data bits, XOR logic operation is also performed with the obtained with logic 1 or 0 value. It depends on the value of the control registers. The parity bit depending on the data bits is calculated by the following formulas:

$$P_{even} = d_{n-1} \oplus \ldots \oplus d_2 \oplus d_1 \oplus d_0 \oplus 0$$
$$P_{even} = d_{n-1} \oplus \ldots \oplus d_2 \oplus d_1 \oplus d_0 \oplus 0 \quad (5.2)$$

In (5.2), P_{even}—parity bit is calculated by logical 0; P_{odd}—parity bit is calculated by logical 1; and d_n is the n-th bit of data.

If a parity bit is active in the packet, it is inserted between the last data bit and the first "end" bit.

Although the UART interface is widely used to transfer data between different devices, it has some drawbacks. The data transfer speed is limited to 1 Mbps and does not have enough flexibility.

Data Transmission Problems Between Clock Domains in High-performance Heterogeneous ICs

Modern heterogeneous ICs have different external interfaces [59]. Each external interface is synchronous, consisting of data bits, control bits, and clock signals [60]. Each interface is formed by its own clock domain. Data transfer from one clock domain to another is characterized by clock domain crossing (CDC) [61].

Data transfer from one domain to another is impossible without matching circuits. Various problems may occur during CDC depending on the parameters of clock signals. Frequency and phase are important parameters of clock signals. The mismatch between two different domains leads to a violation of timing parameters for establishing and maintaining DFFs. Violation of these timing parameters causes a metastable state in DFF output [62].

An example of connection of two flip-flops is presented, where CDC occurs (Fig. 5.16). Here, DFF1 is clocked by clk1 clock signal (CS) and DFF2 by clk2 clock signal.

If clk1 and clk2 signals are synchronous with each other, then all timing parameters will be preserved, and the data In1 will be transmitted losslessly to Out2 of DFF2. However, if clk1 and clk2 signals are asynchronous, TCC occurs, and for a certain period, a violation of the confirmation and maintenance time parameters may occur at Out1 of DFF2.

An example of connecting two DFFs where CDC is generated is presented in Fig. 5.16. Here, DFF1 is clocked by clk1 clock signal (CS) and DFF2 by clk2 clock signal.

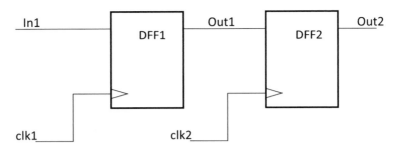

Fig. 5.16 The occurrence of a CDC violation

Fig. 5.17 Occurrence of metastability due to CDC

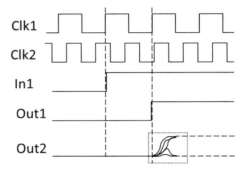

If clk1 and clk2 signals are synchronous towards each other, then all timing parameters will be preserved, and In1 data will be transmitted losslessly to Out2 of DFF2. However, if clk1 and clk2 signals are asynchronous, CDC occurs, and for a certain period, a violation of the confirmation and maintenance timing parameters may occur at Out1 of DFF2.

This will cause a metastable state at the Out2. A timing diagram is presented in Fig. 5.17 where a CDC violation occurs and a metastable state occurs at the Out2 [63].

In case of metastability, the output of the flip-flop is uncertain and can be read as a logic 1 or a logic 0 by the fetching circuit. In addition to functional failure, another problem arises. The duration of the metastable state is also uncertain. Depending on technological parameters used and the frequency of the circuit, this duration can vary from a few percent of the reading CS to the size of a whole paragraph or more. In the latter case, it can cause metastability in the outputs of sequentially connected flip-flops, which also leads to the failure of the entire circuit. There are two concepts to describe CDC: destination and endpoint. The source domain is the one that transmits the data, and the endpoint domain is the one that reads the data. Another problem arises when the CS frequency of the source domain is higher than that of the destination domain. The phenomenon of data loss in case of CDC violation is presented in Fig. 5.18 [64].

5.1 General Issues of Designing Means for High-performance...

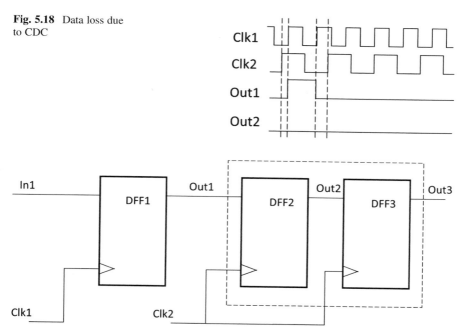

Fig. 5.18 Data loss due to CDC

Fig. 5.19 Structure of a dual-flop synchronizer

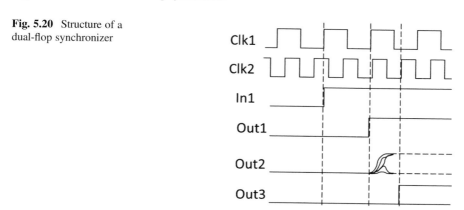

Fig. 5.20 Structure of a dual-flop synchronizer

To solve the above problems and avoid metastability, synchronization circuits are used. A well-known matching circuit is the multi-flop synchronizer. A dual-flop synchronizer is presented in Fig. 5.19 [65].

The operating principle of that circuit is to filter the metastability. As shown in Fig. 5.20, here also a metastable state occurs at the output of DFF2. However, it will not be transferred to the output of clk3 if a logic 1 or a logic 0 is asserted in Out2 at the next positive edge of clk2. The timing diagram of the matching process is presented.

As it can be seen, the Out3 of DFF3 does not go into a metastable state due to CDC and does not cause a behavioral failure of the circuit. If the frequency of clk2 clock signal is high enough, and the duration of metastable state at the output of DFF2 exceeds the duration of the period of clk2, then a metastable state transition is also possible at the output of DFF3. In modern ICs, clock signal frequencies reach several GHz, and in order to avoid CDC problems, three-stage/four-stage synchronizers are used.

There is a formula to calculate the probability of a CDC failure. It describes the time between failures and is a function depending on many variables, including source-destination clock frequency and destination-domain clock frequency. In every IC design problem, it is important to calculate the mean time between failure (MTBF) for an arbitrary signal crossing the CDC boundary. Failure means that the signal passing through the synchronizing flip-flop goes into metastable state and remains in metastable state until the next positive edge of the fetching clock signal. This causes a metastable state at the input of the fetching circuit in the endpoint. The MTBF is calculated by the following formula [66]:

$$\text{MTBF} = \frac{e^{\frac{t_h}{\tau}}}{f_d * f_{cs} * T_m} \qquad (5.3)$$

where t_h is the necessary setup time of data (s), τ time constant of latch (s), f_d data change frequency (Hz), f_{cs} clock frequency of destination domain (Hz), and T_m metastability duration(s).

Implementation Issues of High-performance Heterogeneous IC Architecture

Heterogeneous IC architectures represent one or more CPUs and noncomplex computing cores or accelerators. Among the mentioned architectures, SIMD or MIMD architectures are used in heterogeneous ICs [47]. SISD processors or their grouping are intended to implement non-computational functions or the part of the program that does not need acceleration.

Accelerators are designed to perform computational operations or processes, providing improvements in power consumption or performance parameters as well as operations with big data. When designing heterogeneous architectures, it is important to consider the following: by what kind of interfaces the accelerators and processors are connected to each other, and how the data transfer between them is organized.

One approach to designing heterogeneous architectures is to integrate the accelerator directly into the processor core and transfer data through internal registers. These types of heterogeneous ICs are called application-specific instruction processors (ASIPs). This approach provides acceleration of the implementation of

5.1 General Issues of Designing Means for High-performance... 239

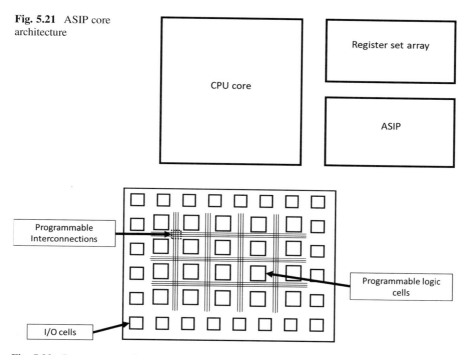

Fig. 5.21 ASIP core architecture

Fig. 5.22 Components and structure of FPGAs

algorithms and low power consumption, but it is not compatible and is used in limited problems. The architecture of ASIP core is shown in Fig. 5.21 [67].

Another design approach is to have accelerators with more complex architectures in ICs that are compatible and capable of performing various computational operations. Here, the processor and accelerators are connected to each other through special packages. The processor is also a control unit for accelerators.

Field programmable gate arrays (FPGAs) also play an important role in modern heterogeneous systems [68]. FPGAs (Fig. 5.22) were originally used for prototyping of digital ICs. However, along with the development of the structure of FPGAs, their fields of application also developed. For example, FPGAs are used in consumer devices. Sometimes, using FPGA in various devices is more affordable and efficient than producing application-specific ICs for these devices, but in terms of operating frequency and power consumption parameters, ICs surpass FPGAs [69].

FPGAs, unlike ICs, are limited by their operating frequency. Modern FPGAs can have an operating frequency of 1–1.5 GHz, and in areas where process parallelization leads to increased performance, FPGAs are widely used.

FPGAs enable programmers to design circuits at a high abstraction level using the OpenCL library [70], which significantly reduces design time. However, with this method, the design is not efficient in terms of resource use and power consumption.

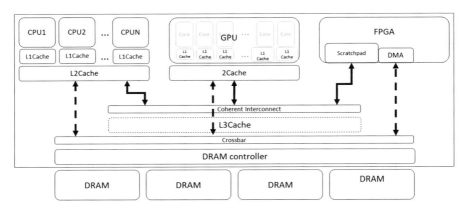

Fig. 5.23 Heterogeneous IC architecture

There are also architectures that are designed on the following principle: data exchange between accelerators and the processor is carried out through a common cache memory. In addition to cache, dynamic memory is also used as shared memory [71].

An example of heterogeneous IC architecture is presented in Fig. 5.23 [38]. At a high level of abstraction, the aforementioned heterogeneous IC consists of CPUs, GPUs, FPGAs, coherent interconnects (CI), last level cache (LLC), dynamic random access memory (DRAM) interconnect, dynamic memory control unit, and external DRAMs. Central processors can consist of 1 or more cores, with a maximum of 16. They, in turn, have L1 cache (1 MB). The cores have a common L2 cache (2 MB).

A GPU consists of hundreds or thousands of computing cores (CPUs). GPUs, like CPUs, have a total of L2 cache. They provide data flow between LLC and processing nodes. Between LLC and L2 cache is CI, which solves the problem of cache memory coherence. In this heterogeneous IC, there is also an embedded FPGA, which makes it possible to perform calculations specific to the given problem. External DRAMs are used to buffer large amounts of data.

Thus, the above-discussed methods ensure the design of heterogeneous ICs, but their application leads to limitations for implementing an even more effective design. In particular, data transfer between component parts is carried out by means of reliable parallel power rails. However, as mentioned in Sect. 5.1.1, the presence of a large number of parallel power rails contributes to the complexity of interconnects between them, which reduces the number of additional cores installed in IC. Clock domain crossing solutions ensure lossless data transfer from one domain to another, but cause additional delays in the process, which negatively affects performance. The structure of heterogeneous ICs discussed above provides high performance, but circuits with such architecture are not compatible and only solve a certain problem.

5.1.3 Principles of Design Means for High-performance Heterogeneous Integrated Circuits

Taking into account the above-discussed architectures of modern heterogeneous ICs and their implementation methods, it can be concluded that they are applicable, but there are certain shortcomings, including:

- Provision of data transfer between components in heterogeneous IC. In the solutions discussed above, data transmission is carried out using parallel power rails, which lead to a high density of power rails at the design stage, which in turn complicates the structure of the entire interconnect. As a result, the number of cores integrated into the IC decreases, which in turn leads to a decrease in performance.
- As a result of the presence of multiple clock domains in heterogeneous ICs, CDC problems arise. In particular, when transferring data from one clock domain to another, a metastable state occurs in the output of the fetching FFs. The solutions discussed above require delaying the multiple clock signal periods by means of synchronizers which contributes to reducing the performance of the entire system.
- The choice and design of heterogeneous IC architectures is unique depending on technical requirements and the field of application. That is, they are not flexible, and their design is a time-consuming process; in addition, creating a verification environment in the process of designing digital systems also requires additional time, which leads to an increase in the cost of designing heterogeneous ICs.

To solve these problems, the following design principles have been proposed for heterogeneous ICs.

- Implementation of data transfer between components in heterogeneous ICs. It is proposed to carry out this process through a modified UART interface, in which the transfer of control data will be carried out, which will ensure flexibility depending on the operating mode. The application of the proposed principle will simplify the structure of interconnects, but will lead to an increase in the area of components.
- Data transfer approach between clock domains. New circuits for synchronizing data transfer between clock domains and corresponding behavioral models have been proposed. Application of the proposed nodes within CDC boundaries will result in additional delay reduction at the expense of additional area consumed.
- Heterogeneous IC architecture. It is proposed to develop such an architecture of heterogeneous IC, which will be structurally and behaviorally compatible, and the design process will be simplified. Namely, the proposed architecture will be generative depending on the technical task. Due to the compatibility of the proposed architecture, the area of heterogeneous IC will increase in comparison to ASIC, with the same performance for solving unique problems.

Conclusions
1. One of the most urgent problems in the creation of modern integrated circuits is the development of means of designing high-performance heterogeneous integrated circuits, which can significantly improve their performance and power consumption and shorten the design period.
2. The main obstacles of the design of high-performance heterogeneous integrated circuits and the organization of corresponding process are the large number of parallel data transfer circuits between cores and the development of data transfer mechanisms between clock domains. An effective way to improve it is the development of means for the design of high-performance heterogeneous integrated circuits.
3. Research and analysis of existing approaches and means of developing means for designing high-performance heterogeneous integrated circuits show that their efficiency does not meet the strict requirements of current market.
4. Approaches to the development of high-performance heterogeneous integrated circuit design tools have been proposed, which allow to significantly improve their main technical parameters: performance, power consumption, and data transfer mechanisms between their components, and reduce the design time.

5.2 Methods of Design for High-performance Heterogeneous Integrated Circuits

5.2.1 Method for Improving Data Transfer Between Components in High-performance Heterogeneous Integrated Circuits

In heterogeneous ICs, depending on the operating mode of cores, one core can have high frequency of the clock signal, the other—low one. These cores have the same data transfer function and interfaces based on the respective instructions. These interfaces complicate the logic function of the interconnect and limit the number of cores in the IC. In order to solve these problems, it is proposed to carry out data transmission between cores through a serial flexible modified UART interface.

As already mentioned, one of the interfaces used for data transfer in ICs is UART, which is widely used due to its simplicity and flexible parameters. UART has transmit and receive lines for data transfer. The applicability of an interface, for example, in computer or network systems, is due to the fact that it is common to many devices, but the data structures change from device to device due to differences in their transfer rates and packet parameters. Due to its flexibility, the interface has a very wide applicability in processor and microprocessor systems, too. The devices in which UART is used are configured so that it is possible to transfer data through them. However, the user needs to implement special settings for satisfactory interface performance depending on the implementation environment. To solve this problem, the UART modification mechanism is proposed, which enables a self-calibration process to be carried out through UART units, as a result of which its parameters will be detected. Since the modified UART was used to transfer data between the internal components of a heterogeneous IC, its performance was significantly increased due to the reduction in physical size of transistors. It would not be possible to implement it for data transfer outside IC. Currently, self-calibration UART controllers exist, but their disadvantage is that only the speed parameter is detected during the self-calibration process. They are not capable of detecting the number of data bits in the packet and the method of calculating the parity bit. A new instruction set is proposed to implement the proposed UART modification.

UART controllers in different devices have different parameters. They are:

- Wide performance range—300 bps - 1 Mbit/s
- 5, 6, 7, and 8 data bits in one packet
- 1, 1.5, and 2—"end" bits
- Calculation of parity bit by odd or even method.

It can be seen from the above that the UART packet can have various structures. Data transfer with UART interface is carried out through one line. This is done by sequential, bit-by-bit shifting. From the perspective of each device, one line is for transmission and one line is for reception. The transmission TX line is connected to the receiver of the other device, and the reception RX line is connected to the transmitter of the other device (Fig. 5.13). The transmitter and receiver lines are independent and asynchronous.

However, devices supporting this interface do not have a dynamic control method, which means that it is not possible to dynamically change the data transfer rate, packet structure, and parity bit calculation method, which affects the flexibility of the interface.

To solve the above-mentioned problems, a modified UART interface control unit (CU), which has the ability of dynamic self-calibration, is proposed [72]. In the process of self-calibration, UART controllers in various devices can detect the data transfer rate of the transmitter, the structure of the packet, and the logic of the parity bit calculation.

The operation of a typical UART is not based on this. In it, both sides of the transmission have the same speed, and the transmission frequencies are independent of each other. The modified UART interface is based on "master-slave" transmission, as in the Universal Serial Bus (USB) interface. The software of the "master" UART unit determines when the self-calibration process should start. It starts with the release signal, which corresponds to the logical value "0" (Fig. 5.24). The duration of this signal depends on the value of the pre-set minimum speed setting. The minimum UART speed is set to 300 bps. For reset, it is necessary to transmit the logical value "0" to the "master" node, equivalent to the duration of 11 bits of time. At a bandwidth of 300 bit/s, the duration of the signal will be 36.6 ms.

"Slave." The control unit detects the reset signal, after which it transmits an 8-bit data packet to the "master" control unit, but the parity bit is missing there. That

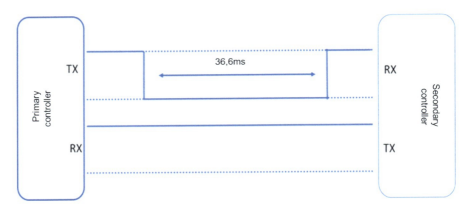

Fig. 5.24 Reset process in modified UART

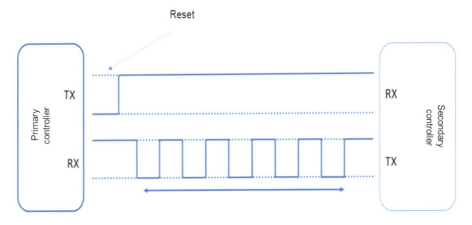

Fig. 5.25 Speed decoding process in modified UART

packet is designed for data bandwidth decoding. The data value in the packet should be 0xAA (0b10101010) (Fig. 5.25). During data reception, the "master" control unit decodes the UART speed parameter of the "slave" control unit. If the "slave" control unit does not respond to the reset signal, then an attempt is made through the "master" control unit to carry out the self-calibration process two more times. After three failed processes, the "master" control unit switches to standard UART operation mode at minimum speed.

After restoring the value of the speed parameter, the "master" control unit performs further transmission accordingly. With this mechanism, the "slave" control unit informs at what speed it can receive a transmitted packet. A new instruction has been announced to detect the rest of parameters for the UART bus transfer. After reset and receiving the packet intended for restoring performance, the "slave" control unit transmits a packet for determining the parameter of the amount of data (Fig. 5.26).

5.2 Methods of Design for High-performance Heterogeneous Integrated Circuits

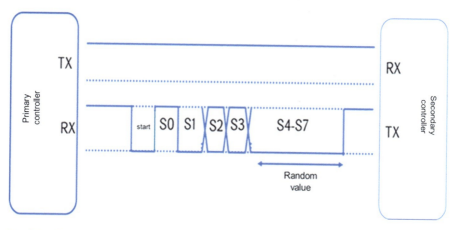

Fig. 5.26 The process of decoding the amount of data in a modified UART

The first two data in the received packets correspond to the instruction identification bits. The corresponding value of the instruction for the amount of data per packet is 2'b01. The next two bits determine the value of the data amount parameter. They are:

- 2'b00—5 bits of data in one packet
- 2'b01—6 bits of data in one packet
- 2'b10—7 bits of data in one packet
- 2'b11—8 bits of data in one packet

After receiving the value of the amount of data parameter, the "slave" control unit passes it to the next packet to determine the number of "end" bits. Its instruction identification value is 2'b10 (Fig. 5.27).

The second and third bits in the received packet determine the number of "end" bits. Their corresponding values are:

- 2'b00—1 end bit
- 2'b01—1.5 end bits
- 2'b10—not applicable
- 2'b11—2 end bits

The parity bit mode is determined in the same way. Its corresponding instruction identification value is 2'b11 (Fig. 5.28). The values corresponding to the second and third bits are:

- 2'b00, 2'b10—the packet does not contain a parity bit.
- 2'b01—even mode of parity bit.
- 2'b11—odd mode of parity bit.

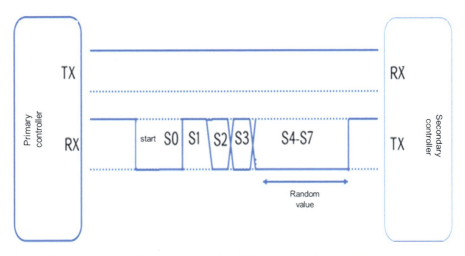

Fig. 5.27 Process of decoding the number of "end" bits in a modified UART

Fig. 5.28 Process of decoding the parity bit mode in a modified UART

After the decoding of all parameter values, the "slave" control unit can decide to end the self-calibration process by sending an instruction, the identification value of which will be 2'b00, or change the values of some parameters by sending respective instructions.

With the mechanism mentioned above, the "slave" control unit transmits the calibration parameters to the "master" control unit. A further extension of the modified UART allows the transmitter to modify the packet structure and speed parameters of the receiver via a special packet. That extension represents the independence of the transmit and receive lines in terms of packet parameters and

5.2 Methods of Design for High-performance Heterogeneous Integrated Circuits

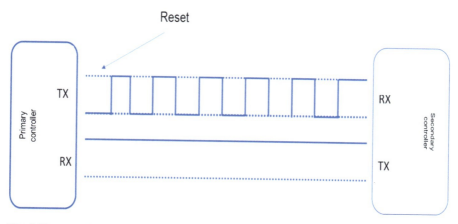

Fig. 5.29 Forced performance change in a modified UART.

transmission speed. That is, the "master" control unit transmits the data at one speed, and the "slave" control unit at another. The same applies to packet parameters.

After the reset, the transmitter immediately transmits the performance calibration packet, based on which the performance parameter is reset in the receiver based on this data (Fig. 5.29). For example, the core is in high-performance mode and transfers data at a speed of 200 Mbps. Before going into low power mode, the core changes the performance setting to 9600 bps by the above mechanism and goes into low power mode. After that, the "slave" control unit restores the data with a bandwidth of 9600 bps. After the reset, the slave control unit's receiver is clocked at the maximum possible frequency, due to which every possible performance parameter is restored (300 bit/s to 200 Mbit/s). Thus, each transmitter can change the performance parameter of the receiver. However, after this process, the transmitter performance of the "slave" control unit does not change, and it can change the corresponding parameter of the "master" control unit with the same mechanism.

Evaluation of the Effectiveness of the Method for Improving the Means of Data Transfer Between Components in High-performance Heterogeneous ICs

In order to evaluate the effectiveness of the method presented in Sect. 5.2.1, a logic circuit of interconnect between two cores was designed. It is implemented by means of adapters between special-purpose clock domains. Then, the same problem was solved using the proposed modified UART, and the area and performance parameters were evaluated.

In order to evaluate the accuracy of the behavioral description of the logic circuit of data transfer between two cores, its testing environment was created (Fig. 5.30). Data processed in heterogeneous ICs is not transferred between cores, or it happens

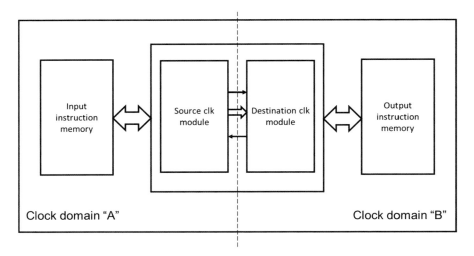

Fig. 5.30 Testing environment of a logic circuit of interconnect

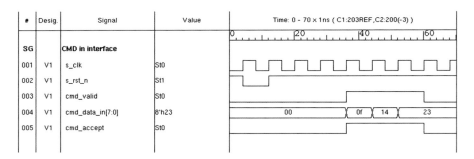

Fig. 5.31 Input interface of a synchronizer

when both cores are in the same power consumption mode. In that case, both cores are clocked with the same clock signal.

However, there is often a transfer of instructions or configuration data from one core to another. The transfer of instructions can occur when the cores are in different power consumption modes and are clocked by asynchronous signals. That is why synchronizers are used.

In this case, the observed circuit implements 8-bit instruction transmission. The data flow is organized in a clock domain from "A" to "B." In addition to 8-bit direct data, control signals are also transmitted and in both directions. Feedback is necessary to transmit the response, which means that the sent instruction has been received at the endpoint "B" clock domain. The instructions transmitted for simulation are stored in the input instruction array and passed to the synchronizer one by one. The interface between them consists of an 8-bit bus, valid, and handshake signals (Fig. 5.31).

5.2 Methods of Design for High-performance Heterogeneous Integrated Circuits

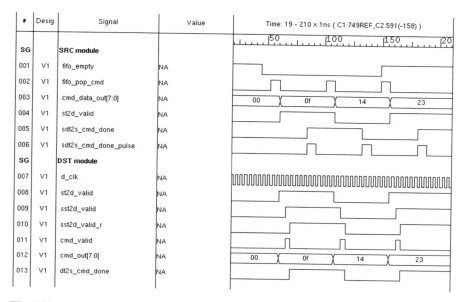

Fig. 5.32 Signals of synchronizers

The source-domain synchronizer can buffer up to eight instructions, at the expense of internal FIFO memory. The command is transmitted with toggle valid signal. As soon as an instruction is read from memory, a switching signal is generated and transmitted to the destination domain. The instruction is considered valid when switching occurs. That signal is transmitted to destination through a sequential synchronizer. The instruction data is not synchronized at the expense of delay time constraint (5.4).

$$T_{\max.\text{delay}} = 2 * T_{\text{dest clk}} - T_{\text{setup}}. \tag{5.4}$$

The acknowledgment signal is transmitted from the destination to the source domain, the synchronization of which is also done by switching (Fig. 5.32).

From the circuit above, it is obvious that eight parallel buses are used to transmit 8-bit data. Below is a solution using a modified UART.

The following changes were made to implement the modified UART:

- To improve the area, provision of packets in different formats has been eliminated. The number of data bits in one packet is 8, the parity bit is absent, and 1 bit is used to transfer "end." In heterogeneous ICs, different cores are in a single IC, and there is no need to support different data packet formats. In contrast to the interconnection outside IC, the influence of external noises on the internal connections in heterogeneous ICs is minimal. That is why there is no need for a parity bit. Redundant internal counters have been removed from the circuit.

- To improve power consumption, clock signal modification interface has been added. Depending on the power consumption mode, the input clock signal changes. For example, in low-power mode, it is blocked or has a low frequency of the order of MHz. For the classical UART implementation, the frequency of the clock signal does not change during operation, and the speed is determined only by maximum values of internal counters. Due to the added interface in the modified UART, it changes according to the frequencies of the cores in IC. Moreover, the number of clock signal frequencies is adjustable.

Detection of performance is implemented when the input clock signal frequency of the circuit is 2 GHz. At maximum performance, the data transfer speed with UART clock is 200 Mb/s (Table 5.2).

For behavioral verification of the receiver, its testing environment was created (Fig. 5.33). It consists of UART transmitter verification intellectual property (VIP), clock signal change control unit, clock signal generator (CSG), and transmitted data receiver (DR).

Table 5.2 The provided performance and clock signal frequencies in the modified UART

Clock signal coding	Clock signal frequency (MHz)	Number of corresponding clock cycles 2 GHz frequency	Speed (MB/s)
00001	1	20,000	0.1
00010	500	40	50
00100	1000	20	100
01000	1500	13.3	150
10000	2000	10	200

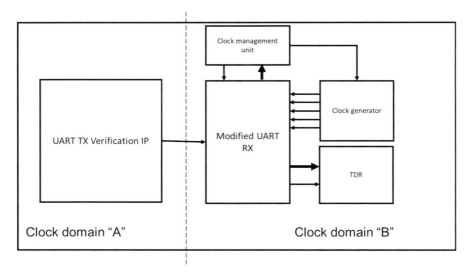

Fig. 5.33 Verification environment of a modified UART receiver

5.2 Methods of Design for High-performance Heterogeneous Integrated Circuits

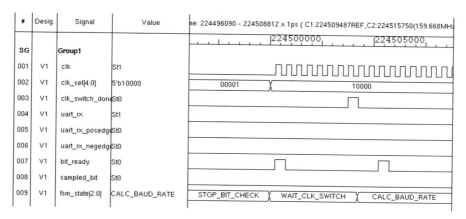

Fig. 5.34 Changing the clock signal in a modified UART

In a modified UART, the receiver's finite state machine (FSM) has two additional states that are anticipated for switching of input clock signal.

Those states are:

- WAIT_CLK_SWITCH—the modified UART receiver waits for the logical one value from validation signal from clock signal change control domain.
- CALC_BAUD_RATE—performance parameter recovery is carried out using a counter clocked at a frequency of 2 GHz.

To switch its input clock, UART receiver modifies its clock select (clk_sel) output and waits for clock switch done (clk_switch_done) signal assertion (Fig. 5.34).

The data rate calculation is performed when receiving the input data 0x55. The duration of one logical 0 bit is calculated using a counter clocked at 2 GHz (Fig. 5.35), after which it again changes the code of the clock signal accordingly (from 200 to 100 Mb/s) and goes to state of waiting for valid/done signal assertion.

The start of the packet is detected by the negative edge of the line (Fig. 5.36).

The receiving bus passes through a two-stage synchronizer, and single clock signals are generated from the positive and negative edges from the received signal.

After that, reception of sequential data begins, and the check of the "end" bit is carried out (Fig. 5.37).

In case of its wrong value, it is accepted as a reset instruction, and the process of changing the clock signal and performance starts (Fig. 5.38). After the change of the clock signal, respective new packets are received (Fig. 5.39).

The transmitter and receiver of the modified UART perform synchronization of clock signals independently of each other (Fig. 5.40).

From the logic synthesis result of the modified UART transmitter (Fig. 5.41), it becomes clear that this circuit can be clocked by a synchronizing signal up to 2.12 GHz.

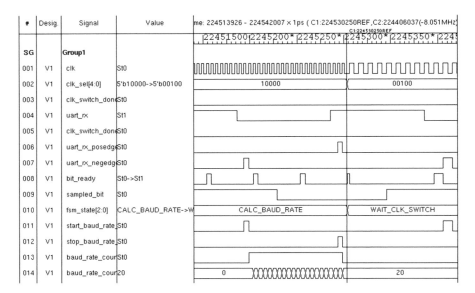

Fig. 5.35 Restoring performance in a modified UART

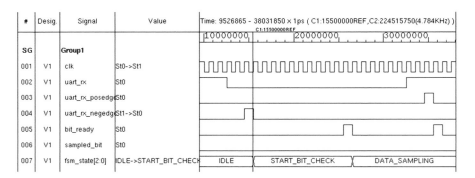

Fig. 5.36 Detecting the start of the packet on a modified UART

The logic synthesis of the receiver showed that its maximum operating frequency is 2.996 GHz (Fig. 5.42).

Thus, the area occupied by the logic circuit of the modified UART is 74.85% more than the area of the core interconnection synchronizer, due to which the number of interconnection buses between the cores has decreased by 8 times (Table 5.3). Compared to the classic UART control unit, the area of the modified UART has decreased by approximately 48%. The increase in the area of the core when using the modified UART was only 2.25%.

5.2 Methods of Design for High-performance Heterogeneous Integrated Circuits

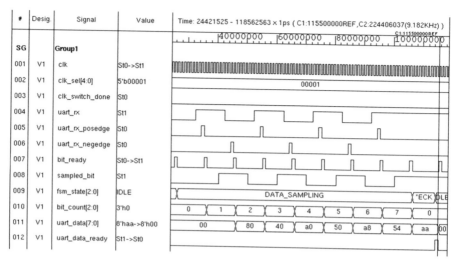

Fig. 5.37 Receiving data in a modified UART

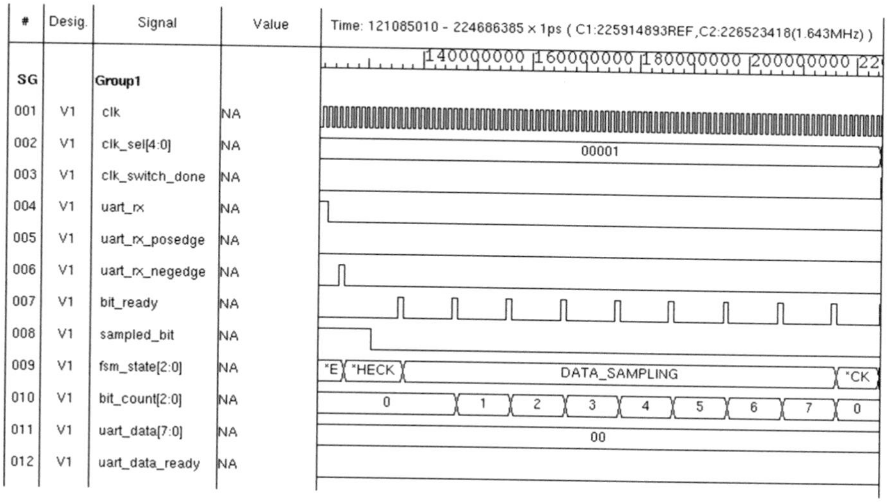

Fig. 5.38 Reset process in a modified UART

5.2.2 Method for Improving Data Transfer Between Clock Domains in High-performance Heterogeneous Integrated Circuits

As it was already mentioned above, in various IC systems, in order to find out how appropriate it is to use a two-stage synchronizer for the solution of the given problem, it is necessary to calculate the MTBF. In case of failure, as a result of

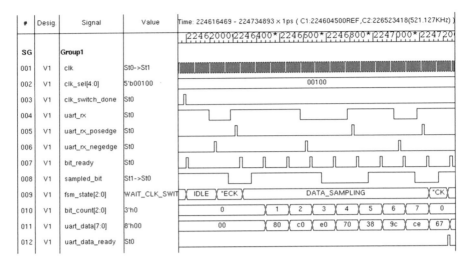

Fig. 5.39 Packet reception with modified performance in a modified UART

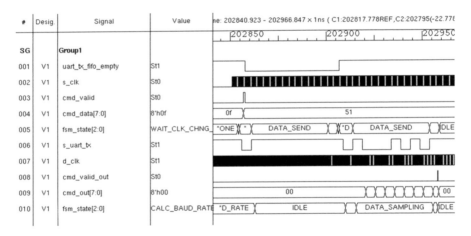

Fig. 5.40 Simulation of transmitter-receiver interconnection in a modified UART

switching of the transmitted signal, a metastable state occurs at the output of the first cascade of the synchronizer. It remains in the metastable state for one complete period of the clock signal and causes a metastable state at the output of the second cascade, too. It can also cause a metastable state in the receiving clock domain. In this case, the behavior of the circuit is violated due to the uncertain values of signals. A variety of factors affect the MTBF, but the clock signal and data change frequencies are decisive.

5.2 Methods of Design for High-performance Heterogeneous Integrated Circuits

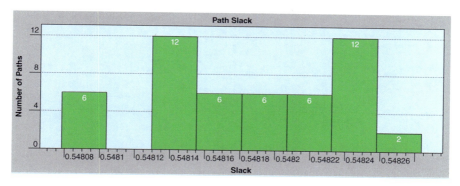

Fig. 5.41 Time margin of a transmitter of a modified UART

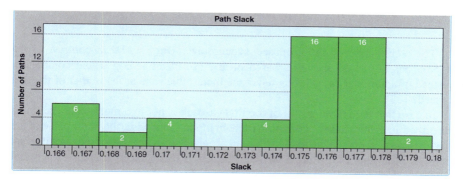

Fig. 5.42 Time margin of a receiver of a modified UART

Table 5.3 Comparison of areas of interconnection circuits

Classical UART	Parallel interconnection circuit	Classical UART	Modified UART
Area unit (μm²)	199.169	677.6	348.26
Number of interconnect buses (units)	8	1	1

$$\mathrm{MTBF} = \frac{1}{f_{c.\,sig.} * f_{data} * X}, \quad (5.5)$$

where $f_{c.\,sig.}$ and f_{data} are the clock signal and data change frequencies, respectively, and X is the coefficient describing the side effects on the MTBF.

Often, in designs clocked at ultra-high frequencies, the MTBF of a two-stage synchronizer is not sufficient to solve the problem, which is why three-cascade or even four-cascade synchronizers are used. However, in those architectures where the occupied area has a decisive role, their use is not advisable.

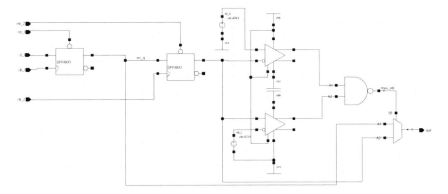

Fig. 5.43 Circuit of a synchronizer

The architecture of the proposed synchronizer (Fig. 5.43), in contrast to the approaches discussed above, is based on the use of an analog signal comparator instead of additional cascades to detect the metastable state at the output of the first cascade [63].

In case that the output signal of the first cascade is not in a metastable state, it is directly transmitted to the destination domain input. Otherwise, the input value of the first cascade is passed instead. The synchronizer with the proposed architecture consists of an input flip-flop, a NAND cell, and a multiplexer. It is also important that the comparators detect the metastable state due to the reference voltages. If the value obtained at the output of the first cascade is in the voltage range of 0.3–0.8 V, the synchronizer considers it metastable. This condition distinguishes which signal the proposed circuit selects and transmits to the output.

Two cases are discussed below: when the signal is stable and when it is in a metastable state.

In the first case, it is assumed that the stable signal has reached the domain. In this case, the output of the comparator will be "01" or "10," depending on the value of the input voltage, greater than 0.8 V or less than 0.3 V, respectively. The role of comparators here is that the value passing through the NAND element, which is the input data of the multiplexer, is a clear logical "1." As a result, the signal from the first clock domain will be transferred to the output.

In the other case, when the value of the received voltage is in the range of 0.3–0.8 V, the output of the comparators will be the logical value "11," which will be transferred to the input of the NAND cell. The latter, in turn, will transfer the value "0" to the input of the multiplexer. As a result, the output of the receiving clock domain will be directly connected to the input, which will make it possible to avoid the transition of a metastable state at the output of flip-flop.

There are strict constraints on certain timing paths of the circuit, which ensure that the fetching flip-flops do not appear in metastable state (5.6). These paths are the output delay of a synchronizer and the input-to-output delay of the first cascade.

5.2 Methods of Design for High-performance Heterogeneous Integrated Circuits

When a metastable state is detected at the output of the first cascade by an analog comparator, the multiplexer transfers the input signal to the output. In that case, when the required delay is not provided, a metastable state can occur in the destination domain at the output of the fetching flip-flop.

$$\begin{cases} T_{\text{in.out.del.}} = T_{\text{max.met.}} \ (T_{\text{max.met.}} > T_{\text{setup}} + T_{\text{hold}}) \\ T_{\text{min.out.del.}} = T_{\text{setup}} + T_{\text{hold}} \ (T_{\text{max.met.}} \leq T_{\text{setup}} + T_{\text{hold}}) \end{cases} \quad (5.6)$$

where $T_{\text{in.out.del.}}$ is the delay from the input to the output of the synchronizer during the switching of the first cascade, $T_{\text{max.met.}}$ is maximum metastable state duration, $T_{\text{min.out.del.}}$ is the minimum output delay, and T_{setup} and T_{hold} are the timing parameters for setup and hold times of fetching flip-flop at destination. Respective time constraints must be given to logic synthesis tool so that the circuit operates without failure in the result.

In digital circuits, a two-stage synchronizer is the basis for the design of synchronizers with complex structures. Based on the proposed circuit, to design complex synchronizers, its behavioral model was proposed in the "Verilog" hardware description language [73]. It is adjustable, so the setup and hold timing parameters of the first cascade which is inside, depending on the library cell, can be changed. It can also change the delay time of output buffers. These timing parameters can be changed randomly within certain limits to detect errors during the simulation.

In case of violation of setup and hold timing parameters, an undefined X value is introduced into the simulation model at the output of the first cascade flip-flop. This mechanism is implemented to simulate metastability. The X value can be passed to read flip-flops during simulation. To simulate this, a random logic value of 1 or 0 is passed to their input. The duration of metastability is also adjustable.

In the proposed model, the necessary timing parameters are embedded, which makes it possible to detect functional errors caused by CDC during the RTL simulation stage and fix them.

Evaluation of the Effectiveness of the Method for Improving Data Transfer Between Clock Domains in High-performance Heterogeneous Integrated Circuits

The two-stage synchronizer (Fig. 5.44) is considered the simplest circuit for synchronizing 1-bit signals. Due to the peculiarity of its structure, the data is delayed by time corresponding to one period. In case that there is a metastable state in the output of the first cascade, then it must be established within the time corresponding to one period, so that there is no metastable state in the output of the second cascade. If the signal is setup, the second cascade will transmit a setup signal with a logic value of 1 or 0 and thereby prevent the transfer of an unstable signal to more sensitive components of IC, the combinational cells. The two-stage synchronizer is a well-

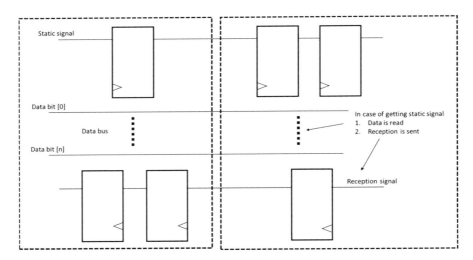

Fig. 5.44 Structure of a feedback synchronizer

known and frequently used circuit in the market. It is not used for fast-to-slow CDCs and can only be used for single-bit data signal synchronization.

A feedback synchronizer is widely used to transfer the entire data stream from one domain to another. As shown above, this problem can be solved by using a two-stage synchronizer for each bit of data stream. However, this method will lead to occupying a large area when there is a 128-bit bus and two additional synchronizers for each bit. In addition to the problem of large circuit area, this method can also lead to data loss in case some bits take the new changed value and the rest take the previous value of data. Instead, having a certain time loss, a synchronizer based on the principle of feedback is used (Fig. 2.21).

Instead of using a two-stage circuit to synchronize each bit of the data bus, the source domain transmits an additional validation signal along with placing the data on the bus. That signal passes through the two-stage synchronizer to the destination domain, and there, upon detecting its active level, the data is read, and the feedback control circuit sends back the receive signal, which is in turn synchronized to the source domain, again through the two-stage synchronizer.

An asynchronous FIFO is used to transfer data from one domain to another asynchronously. Each of domains has its own CS. The source domain and the destination domain receive and read the data with their CSs, respectively (Fig. 5.45).

There are some limitations to consider for this synchronizer. The ratio of read and write speeds should be compatible with the ratio of CS frequencies, so that data loss can be avoided as a result. The FIFO must not go into an overfilled or underfilled state.

The complex synchronizers discussed above are designed using the proposed mixed-signal circuit and compared with classical approaches. "SAED 14 nm"

5.2 Methods of Design for High-performance Heterogeneous Integrated Circuits

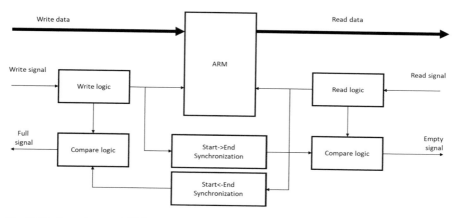

Fig. 5.45 Asynchronous FIFO memory structure

technological library was used [74]. Complex synchronizers are designed using Verilog device description language.

Simulations were performed for two cases discussed above: in the presence of a metastable state of the signal (Fig. 5.46) and in the presence of a stable signal (Fig. 5.47).

In the first case of simulation, the clock signals of receiving clock domains, as well as the transmitted data, are presented. When data is transferred from one domain to the next, it creates a metastable state (v(meta)) (Fig. 2.23), violating the setup and hold times. When metastable signal passes through comparators, a comparison of the current voltage level occurs, outside or within the range of 0.3–0.8 V. In this case, since the condition of metastability occurs, the signal "11" is obtained at the output of comparators, which is transferred to NAND cell. As a result, the multiplexer selection signal changes from 1 to 0. Therefore, the multiplexer connects the input of the domain to the output, which eliminates the problem of generating a metastable state.

When a stable signal (v(meta)) is obtained in the receiving clock domain (Fig. 5.47), the values "01" and "10" corresponding to the stable state are obtained at the output of comparators. In this case, the multiplexer selection signal was received as a logic "1." In this case, the synchronizer designed by the proposed method only transmits the received signal to the output with a certain delay in order to exclude data failure at the destination domain.

From the simulation results of the behavioral model, the metastable state is clearly visible at the output of the first cascade, which is marked as X signal (Fig. 5.48). After the metastable state is completed, its output switches to logic 0 value. After a respective delay, the output of the circuit also switches. As it can be seen, the output of the reading flip-flop in the destination domain is stable. In this case, the input of the synchronizer passes to the output and does not cause a violation of setup and hold timing parameters.

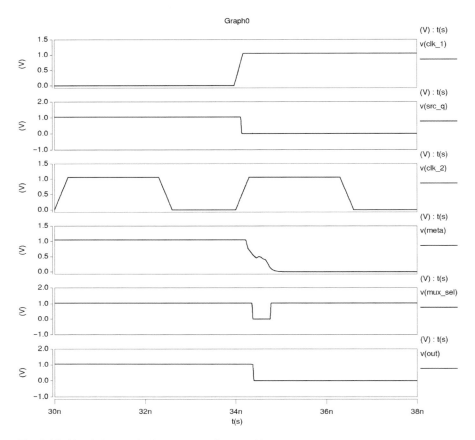

Fig. 5.46 Simulation results by presence of metastable state

In the other, the case is considered when the changed input signal does not pass to the output of the circuit (Fig. 5.49). Here, the output of the first cascade goes into a metastable state, and the multiplexer transmits the data coming from the source domain with a certain delay. As a result, a small jitter occurs. In case that the necessary timing delays are present in the circuit, this is not transmitted to the destination domain at the time of reading, due to which functional failures do not occur. In case that the frequencies of clock signals of source and destination domains are close, the delay of the output of the circuit in case of two consecutive missamplings can be at most two periods.

The designed synchronizer was compared with the classical two-stage circuit (Fig. 5.50). From the simulation result, it can be seen that a time gain of at least one clock signal period was recorded. When using three or more cascades, the number of benefited periods increases, which leads to increased performance in complex IC systems.

5.2 Methods of Design for High-performance Heterogeneous Integrated Circuits 261

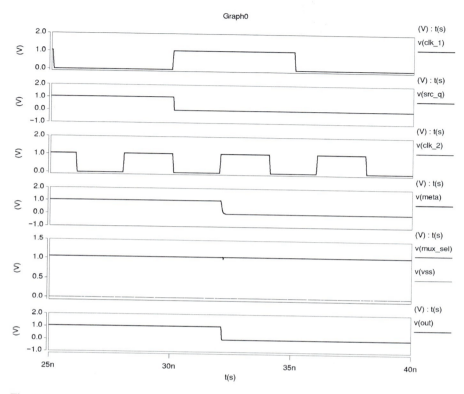

Fig. 5.47 Simulation results with stable signal

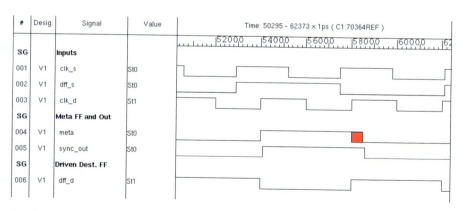

Fig. 5.48 First case of metasteady state simulation of a behavioral model

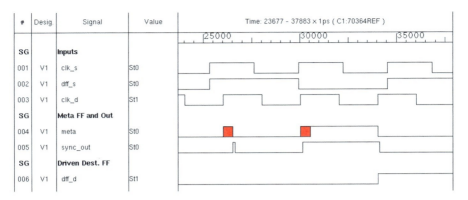

Fig. 5.49 Second case of metasteady state simulation of a behavioral model

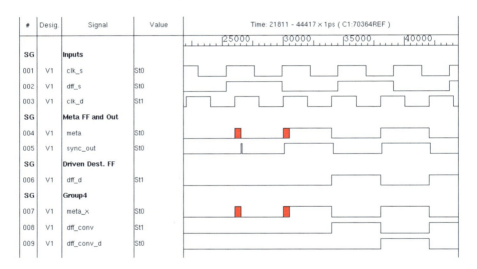

Fig. 5.50 Simulation of the proposed and two-stage synchronizer

A modified pulse signal synchronizer is designed using the proposed mixed-signal synchronization element. It carries out the transmission of the pulse signal generated in the source domain to destination domain. The complexity of the design arises when the pulse signal is transmitted from a clock circuit with a fast frequency to a slow one. Due to the lack of an active edge of the slow clock signal, the signal generated at the source may not be read and will not be generated at the destination domain. That is why feedback is used from the destination domain to the source domain. Due to the small delay of clock signals of the proposed synchronizer, the modified pulse synchronizer transmits the pulse signal at least one period earlier than the classical one (Fig. 5.51).

5.2 Methods of Design for High-performance Heterogeneous Integrated Circuits

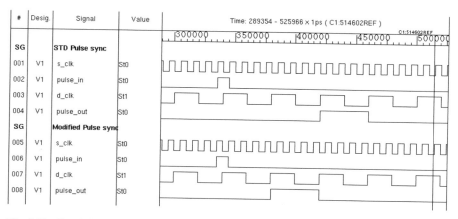

Fig. 5.51 Simulation of modified and classical pulse synchronizers

In addition to the occupied area, the proposed synchronizer has an advantage, that is, when using it, the periods of the additional clock signal are not lost, compared to the method of using cascades. This advantage can be decisive in designs, operating at ultra-high frequencies, where a period of one or two clocks can be used for clock synchronization of other data processing nodes. In the case of the proposed approach, in contrast to the use of cascades, it is possible to reduce the area occupied by the system and increase the performance.

However, in the proposed synchronizer, comparators are used, which are reference voltage-controlled, switching differential amplifiers. That is why the power consumption increases when using the method.

Thus, simulations were performed in the case of a clock signal with a frequency of 2 GHz. At such frequency, in the case of using a three-stage synchronizer, 18 more transistors are used compared to the proposed method. Compared to the two-stage circuit, the proposed circuit occupies an average of 21% more area, but when three or more cascades are used, the area saving is at least 30%.

5.2.3 *Implementation Method of the Architecture in High-performance Heterogeneous Integrated Circuits*

As mentioned above, modern heterogeneous IC implementation methods are complex from the design point of view and inflexible depending on the problem. Therefore, there is a need to create a new architecture of heterogeneous ICs, which, depending on the problem and the field of application, will be adjustable and simple from the point of view of being embedded in other systems.

The architecture with a high abstraction level of the proposed heterogeneous IC consists of (Fig. 5.52):

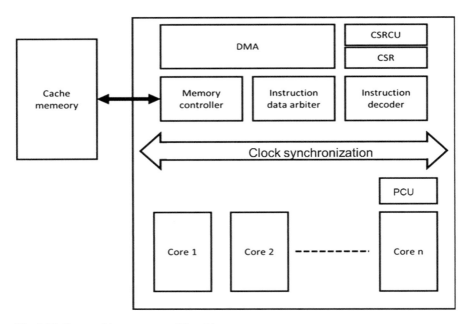

Fig. 5.52 Proposed heterogeneous IC architecture

- Internal cores
- Memory control unit (MCU)
- Queuing unit
- Instruction processor (INP)
- Power consumption control unit (PCU)
- Control/status registers (CSR)
- "Master"—AXI/AHB direct memory access control unit (DMACU)
- "Slave"—AXI/AHB internal register address control unit (IRACU)
- External cache.
- Clock synchronizer module.

The proposed circuit is connected to IC system through the bus of interconnection of peripheral nodes (Fig. 5.53). In a given heterogeneous IC, the CPU executes instructions, sent from the software. It generates unique instructions for the proposed circuit and controls interrupts. It communicates with the internal registers of the circuit with the proposed architecture through the bus of interconnection of peripheral nodes.

Internal cores are computing nodes with RISC architecture. They process 32-bit data. Each CPU has an internal RF which consists of 32 common use registers. Data and instruction memories are separated. The number of internal cores can be from 4 to 16, depending on the circuit creation settings.

The memory control unit is responsible for downloading data and instructions and generating instructions corresponding to the direct memory access node. It also

5.2 Methods of Design for High-performance Heterogeneous Integrated Circuits

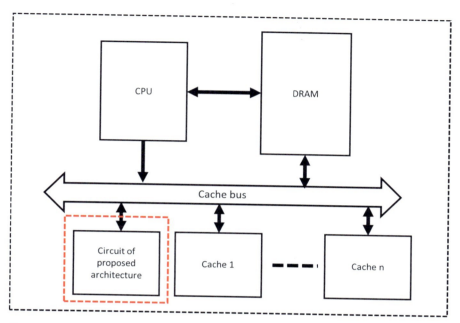

Fig. 5.53 Interconnection of heterogeneous IC in IC system

provides an interface to an external static RAM (SRAM). It cooperates with the queuing unit and special instruction processor provides the process of writing/reading their data.

The queuing unit distributes the implementation of instructions/tasks, coming to an IC. It performs synchronization of CPUs available for data and their processing. In the process of data processing, it performs time allocation of computing resources. Through it, in the case of big data, processing of the latter is performed from an optimal timing perspective. The queuing node generates specific instructions for the CPUs.

The special instruction processor executes the IC-specific instructions that are generated and received by software in the operating system (OS).

The power consumption control unit is responsible for selecting the clock signal frequencies. Depending on the operating mode of the IC, it can switch to low power consumption modes. This unit performs the selection of clock signal frequencies corresponding to these modes.

Control registers are based on flip-flops. They are the control and status registers through which the software in the OS controls IC operation.

"Master" AXI/AHB direct memory access control unit provides data reading or recording from system memory.

"Slave" AXI/AHB register address control unit establishes communication between OS software and control/status registers.

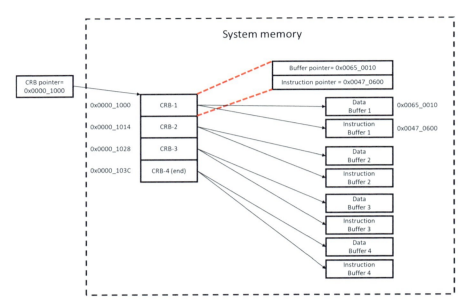

Fig. 5.54 Interrelation of CRB and buffers in system memory

					0x00
	Instruction buffer pointer-low [31:0]				
	Instruction buffer pointer-high [31:0]				0x04
	Data buffer pointer-low [31:0]				0x08
	Data buffer pointer-high [31:0]				0x0C
Instruction buffer size[12:0]	Data buffer size[15:0]	Interrupt	Next	Last	Chain	0x10

Fig. 5.55 CRB structure

External cache memory is for instruction/data buffering. Those domains are separate.

The entire operation of a circuit is controlled by a specially designed software (SDS) in the OS (Fig. 5.54). It prepares data in system memory and generates computing request blocks (CRBs). Each CRB contains a system memory pointer to a block of data and instructions to be processed, as well as four control bits (Fig. 5.55). As it can be seen, each CRB can deal with a 64 KB data buffer and a 4 KB instruction buffer. Each core has an area in cache memory of appropriate size, given to instructions and data.

In order to reduce the load on the cache memory, three external static memory devices are used.

The first one is designed for high abstraction level IC units to buffer and process CRBs. Its size is adjustable and determined by the parameters selected by the user. The second and third ones are the instruction and data buffers, respectively, the sizes of which are determined by the number of cores:

5.2 Methods of Design for High-performance Heterogeneous Integrated Circuits

$$\text{Ins.buf} = N_{\text{core}} * 4 \text{ KB}$$
$$\text{data.buf} = N_{\text{core}} * 64 \text{ KB} \tag{5.7}$$

Before the circuit can start its operation, it must go through the initialization process, which is issued by SDS in the form of instructions. These instructions are processed and implemented by the INS. During this process, internal registers and core resources are adjusted.

After the initialization process is completed, SDS prepares the CRBs and issues their reading and execution process with appropriate instructions, updating the CRB's internal pointer. In order to read and process the CRB, it is necessary that chain bit has logical 1 value. Otherwise, that CRB is closed by the device and ends. The SDS is informed about the status of the CRB through the interrupt mechanism implemented in the circuit. It is used if the interrupt bit in the corresponding CRB has a logic 1 value. Otherwise, the SDS should periodically read the CRB. After completing its processing, the circuit registers a logic 0 value in the chain field.

Buffers of different domains can be connected to each other by registering a logical 1 value in the next field.

Evaluation of Effectiveness of Implementation Method of the Architecture of High-performance Heterogeneous ICs

To evaluate the performance of the proposed architecture, a video processing platform on a high-performance FPGA was designed. Video processing was performed on that platform, and its performance compared to video resolution was studied. The same problem was solved using the proposed heterogeneous IC, and its effectiveness was compared. In case of FPGA implementation, dynamic memory is used to buffer video image data, and data is recorded there with a specially designed interface (Fig. 5.56) [75].

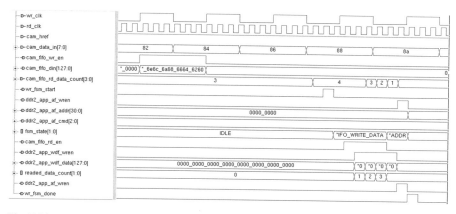

Fig. 5.56 Image registration in external dynamic memory

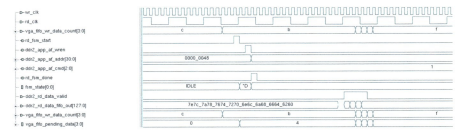

Fig. 5.57 Reading an image from external dynamic memory

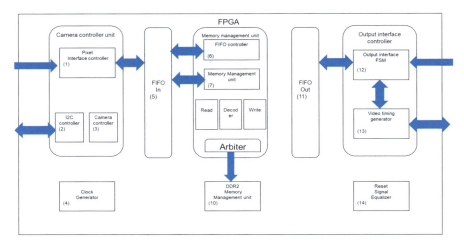

Fig. 5.58 Architecture of a video image processing circuit designed on FPGA

After the full image data is received in the read buffer, it is read and processed (Fig. 5.57).

After reading, the image is transformed from the red-green-blue (RGB) color range to the color-saturation-brightness (CSB) range, undergoes all the appropriate processing and corrections there, and is converted back to the RGB range. Then it is displayed on the monitor.

From the structure of the designed system, it can be seen that it has four clock domains (Fig. 5.58). They are the corresponding domain of the clock signal received from the image reading interface from the camera, the domain of the external DRAM, the domain of the output image extraction interface, and the domain of the reference clock signal.

From the results of the research, it can be seen that the operating frequency of the circuit increases with the increase of the resolution of the video resolution (Fig. 5.59).

In case of maximum frequency, it is possible to provide 60 Hz video processing with a resolution of 1920 × 1200 (Fig. 5.60). Research shows that at an operating

5.2 Methods of Design for High-performance Heterogeneous Integrated Circuits

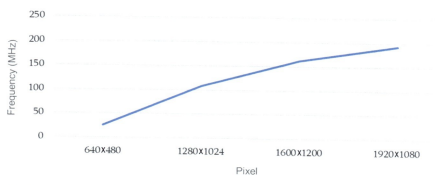

Fig. 5.59 Dependence of frequency and image resolution

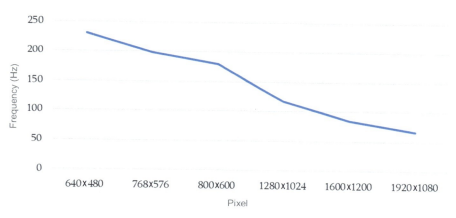

Fig. 5.60 Dependence of video frequency and supporting resolution

frequency of approximately 200 MHz, the designed circuit is able to provide the same amount of data processing as a 60 Hz video stream. Along with reducing the resolution of the video image, the frequency of the possible supporting video stream increases.

The same data processing algorithm is implemented using the circuit with the proposed architecture. However, there are certain differences in terms of receiving data and extracting processed data. In the case of heterogeneous ICs, the image data is located in the system memory and there is no need for additional buffering. They are read using DMACU (Fig. 5.61). After reading data, instructions are read for cores with the same mechanism. After processing data, they are recorded in the system memory (Fig. 5.39). In addition to reading instructions and data, the same process is carried out for CRBs by means of DMACU (Fig. 5.62).

The CRB is read and buffered in the appropriate range in external cache memory.

This is then read from there by the queue unit, which in turn decodes and generates instructions for the DMACU to read the array of instructions and data from system memory (Fig. 5.63).

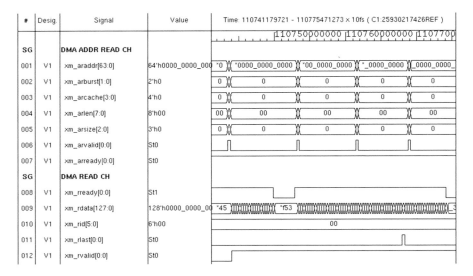

Fig. 5.61 Data reading through DMACU

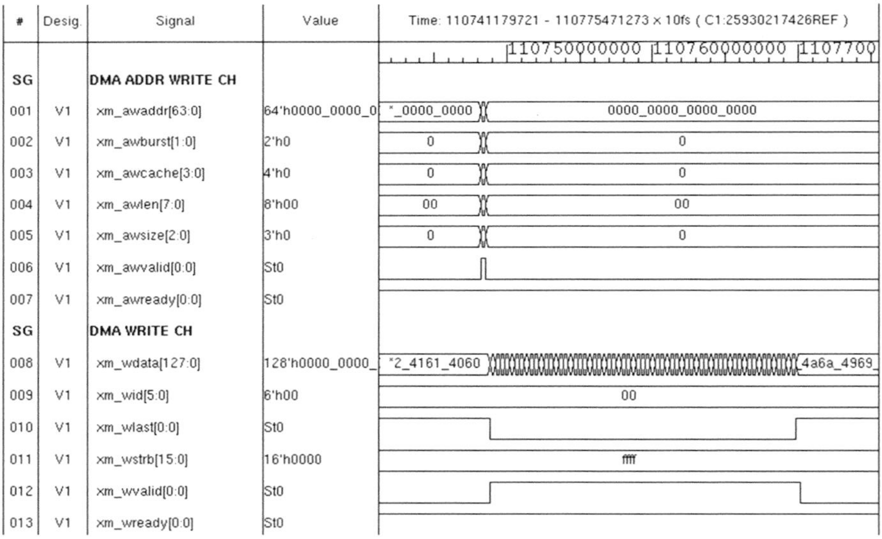

Fig. 5.62 Post-recording of data via DMACU

In order to perform post-recording of data, it is read from the data area of the static memory and transferred to the DMACU through the memory control unit (MCU) (Fig. 5.64). It, in turn, implements the data post-recording process.

The circuit is designed for a maximum frequency of 2 GHz. It has the ability to configure and select the number of cores from 4 to 16. There are strict requirements

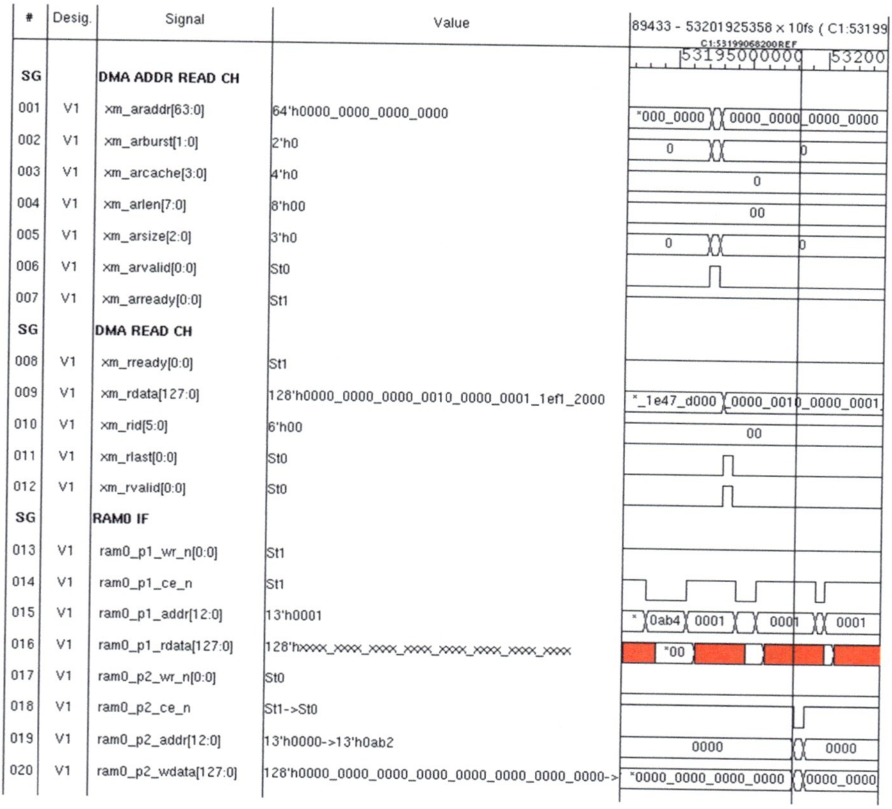

5.2 Methods of Design for High-performance Heterogeneous Integrated Circuits

#	Desig.	Signal	Value
SG		DMA ADDR READ CH	
001	V1	xm_araddr[63:0]	64'h0000_0000_0000_0000
002	V1	xm_arburst[1:0]	2'h0
003	V1	xm_arcache[3:0]	4'h0
004	V1	xm_arlen[7:0]	8'h00
005	V1	xm_arsize[2:0]	3'h0
006	V1	xm_arvalid[0:0]	St0
007	V1	xm_arready[0:0]	St1
SG		DMA READ CH	
008	V1	xm_rready[0:0]	St1
009	V1	xm_rdata[127:0]	128'h0000_0000_0000_0010_0000_0001_1ef1_2000
010	V1	xm_rid[5:0]	6'h00
011	V1	xm_rlast[0:0]	St0
012	V1	xm_rvalid[0:0]	St0
SG		RAM0 IF	
013	V1	ram0_p1_wr_n[0:0]	St1
014	V1	ram0_p1_ce_n	St1
015	V1	ram0_p1_addr[12:0]	13'h0001
016	V1	ram0_p1_rdata[127:0]	128'hxxxx_xxxx_xxxx_xxxx_xxxx_xxxx_xxxx_xxxx
017	V1	ram0_p2_wr_n[0:0]	St0
018	V1	ram0_p2_ce_n	St1->St0
019	V1	ram0_p2_addr[12:0]	13'h0000->13'h0ab2
020	V1	ram0_p2_wdata[127:0]	128'h0000_0000_0000_0000_0000_0000_0000_0000->

Fig. 5.63 CRB recording, read in cache

for the frequency of MCU and DMACU. Their bandwidth should be equal to the sum bandwidth of the cores. Otherwise, the waiting process in the core arbitration process will take a long time, which will lead to a decrease in performance. The number of gates used in one core is approximately 24,000. The area of the total circuit is approximately 210,000 gates.

From the results of logical synthesis of the circuit (Fig. 5.65), the dependence of the number of cores and the used gates can be seen. The dependence is almost linear due to the presence of extra-core units. They occupy almost four core areas. However, in case of further scaling, the area of cores will dominate compared to them.

From the comparison it can be seen that the video image processing circuit, designed on FPGA, is close to the proposed circuit in the case of four-core configuration (Fig. 5.66). In all other cases, the increase in area is multiple, even compared to the initial area (Table 5.4).

From the performance results it can be seen that the performance increases with the increase in the number of cores (Fig. 5.44).

272 5 Design of High-performance Heterogeneous Integrated Circuits

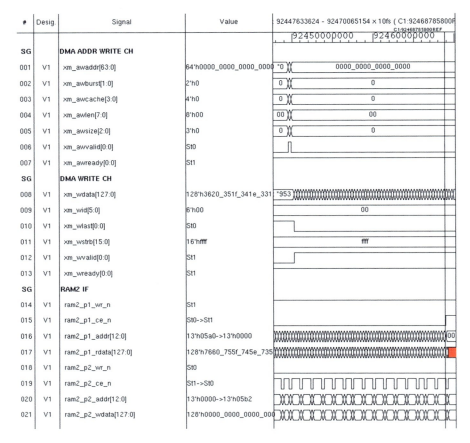

Fig. 5.64 The process of post-recording processed data

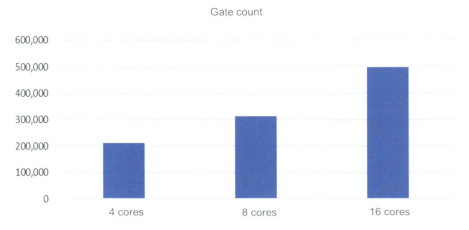

Fig. 5.65 Dependence on the number of cores and the number of used gates

5.2 Methods of Design for High-performance Heterogeneous Integrated Circuits

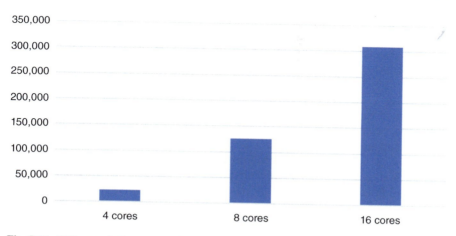

Fig. 5.66 Difference in the number of gates between FPGA and proposed circuits

Table 5.4 Comparison of design results on FPGA and the proposed circuit

	Circuit implemented on FPGA	Proposed 4 cores	Proposed 8 cores	Proposed 16 cores
Number of gates	188,249	211,536	314,147	497,458
Difference	–	23,287	125,898	309,209
Video frame rate (Hz)	~60	~79	~120	~143

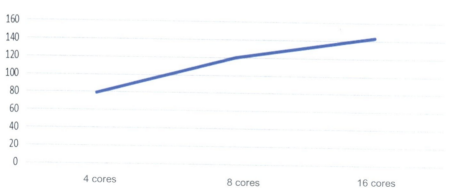

Fig. 5.67 Dependence of the performance of the proposed circuit and the number of cores

As it can be seen from the results, a further increase in the number of cores leads to a sharp increase in area, but the increase in performance is not relevant (Fig. 5.67).

Thus, a method for the implementation of the architecture of heterogeneous ICs was created, which, due to the use of queuing, memory control, and direct memory waiting units and a special instruction, provides a 32.48% increase in speed in the video image processing process at the expense of an 11.008% increase in area.

Conclusions

1. A method of improving the means of data transfer between components in heterogeneous integrated circuits has been proposed, which, due to the modified architecture, provides an eightfold reduction in the number of interconnect buses between cores, at the expense of an increase in the area spent in the core by 2.25%.
2. A method for improving the means of data transfer between clock domains in heterogeneous integrated circuits has been proposed, which, due to the mixed-signal architecture, provides an increase in performance of at least 50% at the expense of an average of 21% additional area.
3. A method of implementing the architecture of heterogeneous integrated circuits was proposed, which, due to the use of queuing, memory control and direct memory waiting units and a special instruction, provides a 32.48% increase in performance at the expense of an 11.008% increase in area.

References

1. Y. Li, X. Zhao, T. Cheng, Heterogeneous computing platform based on CPU+FPGA and working modes. 2016 12th International conference on computational intelligence and security (CIS) (2016), pp. 669–672
2. K. Rupp, Microprocessor trend data (2022). https://github.com/karlrupp/microprocessor-trend-data/tree/master/50yrs
3. M. Gianfagna, What is Moore's law? (2021), https://www.synopsys.com/glossary/what-is-moores-law.html#:~:text=Definition,as E %3D mc2)
4. M.H. Scaling, Power, and the future of CMOS technology. Device research conference (2008), pp. 7–8
5. F. Juan, F. Qingwen, H. Xiaoting, et al., Performance optimization by dynamically altering cache replacement algorithm in CPU-GPU heterogeneous multi-core architecture. 2017 17th IEEE/ACM international symposium on cluster, cloud and grid computing (CCGRID) (2017), pp. 723–726
6. S. Vijayalakshmi, A. Alagan, D.P. Kothari, Power-performance of multi-threaded multi-core processor: analysis, optimization and simulation. 2013 international conference on high performance computing & simulation (HPCS) (2013), pp. 674–677
7. M. Diogo, D. Helder, S. Leonel, I. Aleksandar, Analyzing performance of multi-cores and applications with cache-aware Roofline Model. 2017 international conference on high performance computing & simulation (HPCS) (2017), pp. 933–934
8. R. Ritesh, K. Neeharika, R. Nitin, Digital image processing through parallel computing in single-core and multi-core systems using MATLAB. 2017 2nd IEEE international conference on recent trends in electronics, information & communication technology (RTEICT) (2017), pp. 462–465
9. L. Duk Hyung, C. Hyun Hak, J. Ok Hyun. Analysis of power, temperature, and performance on mobile application processor. International conference on mechatronics, robotics and systems engineering (MoRSE) (2019), pp. 81–85
10. W. Siqi, A. Gayathri, M. Tulika, OPTiC: Optimizing collaborative CPU–GPU computing on mobile devices with thermal constraints. IEEE Trans. Comput.-Aided Des. Integr. Circuits Syst. **38**(3), 393–406 (2019)

11. Jayant, V. Shahi, C.M. Velpula, CPU temperature aware scheduler a study on incorporating temperature data for CPU scheduling decisions. 2015 international conference on advances in computing, communications and informatics (ICACCI) (2015), pp. 2409–2413
12. 2021 Trends. https://static1.squarespace.com/static/6130ef779c7a2574bd4b8888/t/61 6c79ed5a30e36825f47818/1634499069232/isscc2021.press_kit_110620.pdf. Institute of Electrical and Electronics Engineers – University of Pennsylvania (2021), pp. 1–152
13. B. Shekhar, C.A. Andrew, The future of microprocessors. Commun. ACM **54**(5), 67–77 (2011)
14. B. Shekhar, *Thousand Core Chips—A Technology Perspective* (Intel Corp, Microprocessor Technology Lab, Hillsboro, 2012), pp. 746–749
15. White Paper, Next leap in microprocessor architecture: Intel® Core™ duo processor (2006), p. 4
16. A.R.A. Saif, K. Bin Jumari, Performance study of Core2Duo desktop processors. 2009 International conference on electrical engineering and informatics (2009), pp. 532–536
17. M.D. Hill, Amdahl's law in the multicore era. 2008 IEEE 14th international symposium on high performance computer architecture (2008) vol. 41, no. 7, pp. 33–38
18. B. Rubén, B. Daniele, B. Andrea, A. Giovanni, et al., A synchronization-based hybrid-memory multi-core architecture for energy-efficient biomedical signal processing. IEEE Trans. Comput. **66**(4), 575–585 (2017)
19. K. Takanori, L. Yamin, A cost and performance analytical model for large-scale on-chip interconnection networks. 2016 4th international symposium on computing and networking (CANDAR) (2016), pp. 447–450
20. M.J. Cade, A. Qasem, Balancing locality and parallelism on shared-cache mulit-core systems. 2009 11th IEEE international conference on high performance computing and communications (HPCC 2009) (2009), pp. 188–195. https://doi.org/10.1109/HPCC.2009.61
21. J. Ma, C. Hao, W. Zhang, T. Yoshimura, Power-efficient partitioning and cluster generation design for application-specific network-on-chip. 2016 international SoC design conference: smart SoC for intelligent things (ISOCC) (2016), pp. 83–84. https://doi.org/10.1109/ISOCC. 2016.7799744
22. K. Onur, N. Nachiappan Chidambaram, J. Adwait, A. Rachata, Managing GPU concurrency in heterogeneous architectures. 2014 47th annual IEEE/ACM international symposium on microarchitecture (2014), pp. 114–126
23. J. Choquette, W. Gandhi, O. Giroux, et al., NVIDIA A100 tensor Core GPU: Performance and innovation. IEEE Micro. **41**(2), 29–35 (2021). https://doi.org/10.1109/MM.2021.3061394
24. F.L. Yuan, C.C. Wang, T.H. Yu, D. Marković, A multi-granularity FPGA with hierarchical interconnects for efficient and flexible Mobile computing. IEEE J. Solid State Circuits **50**(1), 137–149 (2015). https://doi.org/10.1109/JSSC.2014.2372034
25. Z. Lai, K.T. Lam, C.L. Wang, J. Su, A power modelling approach for many-core architectures. Proceedings of the 2014 10th international conference on semantics, knowledge and grids (SKG-2014) (2014), pp. 128–132. https://doi.org/10.1109/SKG.2014.10
26. F. Conti, C. Pilkington, A. Marongiu, L. Benini, He-P2012: Architectural heterogeneity exploration on a scalable many-core platform. Proceedings of the ACM great lakes symposium on VLSI, (GLSVLSI) (2014), pp. 231–232. https://doi.org/10.1145/2591513.2591553
27. W.P. Huang, R.C.C. Cheung, H. Yan, An efficient application specific instruction set processor (ASIP) for tensor computation. Proceedings of the international conference on application-specific systems, architectures and processors, vol. 2019 (2019), p. 37. https://doi.org/10.1109/ ASAP.2019.00-36
28. H. Anwar, M. Daneshtalab, M. Ebrahimi, M. Ramirez, et al Integration of AES on heterogeneous many-core system. Proceedings of the 2014 22nd euromicro international conference on parallel, distributed, and network-based processing, (PDP 2014) (2014), pp. 424–427. https:// doi.org/10.1109/PDP.2014.86
29. H.-J. Wunderlich, Simulation on reconfigurable heterogeneous computer architectures (2017), https://www.iti.uni-stuttgart.de/en/chairs/ca/projects/oldprojects/simtech/

30. A.Z. Adamov, Computation model of data intensive computing with MapReduce. Proceedings of the 14th IEEE international conference on application of information and communication technologies (AICT-2020) (2020), pp. 1–5. https://doi.org/10.1109/AICT50176.2020.9368841
31. M. Davari, A. Ros, E. Hagersten, S. Kaxiras, An efficient, self-contained, on-chip directory: DIR1-SISD. Parallel architectures and compilation techniques – Conference proceedings (PACT) (2015), pp. 317–330. https://doi.org/10.1109/PACT.2015.23
32. I. Yamazaki, J. Kurzak, P. Luszczek, J. Dongarra, Design and implementation of a large scale tree-based QR decomposition using a 3D virtual systolic array and a lightweight runtime. Proceedings of the IEEE 28th international parallel and distributed processing symposium workshops (IPDPSW-2014) (2014), pp. 1495–1504. https://doi.org/10.1109/IPDPSW.2014.167
33. M.T. Sim, Q. Yi, An adaptive multitasking superscalar processor. 2019 IEEE 5th International conference on computer and communications (ICCC 2019) (2019), pp. 1293–1299. https://doi.org/10.1109/ICCC47050.2019.9064185
34. S. Processors, Superscalar processor: Intro (1995). No. 7, pp. 1–19. https://en.wikipedia.org/wiki/Superscalar_processor
35. SISD, SIMD, MISD, MIMD. https://learnlearn.uk/alevelcs/sisd-simd-misd-mimd/
36. J. Chen, C. Yang, Optimizing SIMD parallel computation with non-consecutive array access in inline SSE assembly language. Proceedings of the 2012 5th international conference on intelligent computation technology and automation (ICICTA-2012) (2012), pp. 254–257. https://doi.org/10.1109/ICICTA.2012.70
37. B.S. Mahmood, M.A.A. Jbaar, Design and implementation of SIMD vector processor on FPGA. 2011 4th international symposium on innovation in information and communication technology (ISIICT'2011) (2011), pp. 124–130. https://doi.org/10.1109/ISIICT.2011.6149607
38. L. Juan Gómez, M. Onur, P&S heterogeneous systems SIMD processing and GPUs (2021), pp. 1–75. https://safari.ethz.ch/projects_and_seminars/fall2021/lib/exe/fetch.php?media=p_s-hetsys-fs2021-meeting2-aftermeeting.pdf
39. B. Rajeshwari, K. Veena, MIMO receiver and decoder using vector processor. Proceedings/TENCON IEEE region 10 annual international conference: 2017, vol. 2017-December, pp. 1225–1230. https://doi.org/10.1109/TENCON.2017.8228044
40. K. Patsidis, C. Nicopoulos, G.C. Sirakoulis, G. Dimitrakopoulos, RISC-V2: A scalable RISC-V vector processor. Proceedings of the IEEE international symposium on circuits and systems, October (2020), pp. 1–5. https://doi.org/10.1109/iscas45731.2020.9181071
41. Y. Xiao, Z. Chen, L. Zhang, Accelerated CT reconstruction using GPU SIMD parallel computing with bilinear warping method. 2009 1st international conference on information science and engineering (ICISE-2009) (2009), pp. 95–98. https://doi.org/10.1109/ICISE.2009.203
42. A. Halaas, B. Svingen, M. Nedland, P. Sætrom, et al., A recursive MISD architecture for pattern matching. IEEE Trans. Very Large Scale Integr. Syst. **12**(7), 727–734 (2004). https://doi.org/10.1109/TVLSI.2004.830918
43. A. Yazdanbakhsh, K. Samadi, N.S. Kim, H. Esmaeilzadeh, GANAX: A unified MIMD-SIMD acceleration for generative adversarial networks. Proceedings of the international symposium on computer architecture (2018), pp. 650–661. https://doi.org/10.1109/ISCA.2018.00060
44. S. Arrabi, D. Moore, L. Wang, K. Skadron, et al., Flexibility and circuit overheads in reconfigurable sIMD/MIMD systems. Proceedings of the 2014 IEEE 22nd international symposium on field-programmable custom computing machines (FCCM 2014) (2014), p. 236. https://doi.org/10.1109/FCCM.2014.71
45. Y. Yamato, N. Hoshikawa, H. Noguchi, et al., A study to optimize heterogeneous resources for open IoT. Proceedings of the 2017 5th international symposium on computing and networking (CANDAR-2017), January (2018), pp. 609–611. https://doi.org/10.1109/CANDAR.2017.16
46. K. Gai, L. Qiu, H. Zhao, M. Qiu, Cost-aware multimedia data allocation for heterogeneous memory using genetic algorithm in cloud computing. IEEE Trans. Cloud Comput. **8**(4), 1212–1222 (2020). https://doi.org/10.1109/TCC.2016.2594172

References

47. A.R. Brodtkorb, C. Dyken, T.R. Hagen, et al., State-of-the-art in heterogeneous computing. Sci. Program. **18**(1), 1–33 (2010). https://doi.org/10.3233/SPR-2009-0296
48. K. Zhu, Y. Ding, Research on low power scheduling of heterogeneous multi core mission based on genetic algorithm. Proceedings of the 9th international conference on measuring technology and mechatronics automation (ICMTMA-2017) (2017), pp. 219–223. https://doi.org/10.1109/ICMTMA.2017.0059
49. C. Yu, M. Cai, An image depth processing method based on parallel computing and multi-GPU. Proceedings of the 2nd international conference on smart electronics and communication (ICOSEC-2021) (2021), pp. 1009–1012. https://doi.org/10.1109/ICOSEC51865.2021.9591686
50. A.K. Gupta, A. Raman, N. Kumar, R. Ranjan, Design and implementation of high-speed universal asynchronous receiver and transmitter (UART). 2020 7th international conference on signal processing and integrated networks (SPIN-2020) (2020), pp. 295–300. https://doi.org/10.1109/SPIN48934.2020.9070856
51. S. Harutyunyan, T. Kaplanyan, A. Kirakosyan, H. Khachatryan, Configurable verification IP for UART. 2020 IEEE 40th international conference on electronics and nanotechnology (ELNANO) (2020), pp. 234–237
52. T. Praveen Blessington, B. Bhanu Murthy, G.V. Ganesh, T.S.R. Prasad, Optimal implementation of UART-SPI interface in SoC. 2012 international conference on devices, circuits and systems, ICDCS 2012 (2012), pp. 673–677. https://doi.org/10.1109/ICDCSyst.2012.6188657
53. V. Melikyan, S. Harutyunyan, A. Kirakosyan, T. Kaplanyan, UVM verification IP for AXI. 2021 IEEE east-west design and test symposium, (EWDTS-2021) (2021), pp. 1–4. https://doi.org/10.1109/EWDTS52692.2021.9580997
54. J. Liu, M. Hong, K. Do, J.Y. Choi, et al. Clock domain crossing aware sequential clock gating. Design, automation & test in Europe conference & exhibition (DATE) (2015), pp. 1–6
55. S. Hatture, S. Dhage, Open loop and closed loop solution for clock domain crossing faults. Global conference on communication technologies (GCCT-2015) (2015), pp. 645–649. https://doi.org/10.1109/GCCT.2015.7342741
56. D. Basu, D.K. Kole, H. Rahaman, Implementation of AES algorithm in UART module for secured data transfer. Proceedings of 2012 international conference on advances in computing and communications (ICACC-2012) (2012), pp. 142–145. https://doi.org/10.1109/ICACC.2012.32
57. B. Zhang, K. Zhang, J. Zhu, X. Li, UART interface design based on DM642 video surveillance system and wireless network module. Proceedings of 2011 IEEE 2nd international conference on software engineering and service science (ICSESS-2011) (2011), pp. 477–480. https://doi.org/10.1109/ICSESS.2011.5982357
58. KeyStone architecture: Universal asynchronous receiver/transmitter (UART). Texas Instruments (2010), pp. 1–51
59. J.H. Hong, S.W. Han, E.Y. Chung, A RAM cache approach using host memory buffer of the NVMe interface. International SoC design conference: Smart SoC for intelligent things (ISOCC-2016). (2016), pp. 109–110. https://doi.org/10.1109/ISOCC.2016.7799757
60. D. Akash, M. Kishore, Mohana, K.H. Basha, Interfacing of flash memory and DDR3 RAM memory with Kintex 7 FPGA board. Proceedings of the 2nd IEEE international conference on recent trends in electronics, information and communication technology (RTEICT-2017) proceedings, January (2017), pp. 2006–2010. https://doi.org/10.1109/RTEICT.2017.8256950
61. S. Zhou, T. Zhang, Y. Yang, cross clock domain signal research based on dynamic motivation model. Proceedings of the 4th international conference on dependable systems and their applications. (DSA-2017), January (2017), p. 156. https://doi.org/10.1109/DSA.2017.34
62. N. Karimi, K. Chakrabarty, Detection, diagnosis, and recovery from clock-domain crossing failures in multiclock SoCs. IEEE Trans. Comput.-Aided Design Integra. Circuits Syst. **32**(9), 1395–1408 (2013). https://doi.org/10.1109/TCAD.2013.2255127
63. V. Melikyan, S. Harutyunyan, T. Kaplanyan, A. Kirakosyan, et al., Design and verification of novel sync cell. Proceedings of the 2021 IEEE east-west design and test symposium, (EWDTS-2021) (2021). pp. 1–5. https://doi.org/10.1109/EWDTS52692.2021.9580985

64. C.E. Cummings, Clock domain crossing (CDC) design & verification techniques using system Verilog. Techniques (2008), No. Cdc. pp. 1–56
65. M. Bartík, Clock domain crossing – An advanced course for future digital design engineers. Proceedings of the 2018 7th mediterranean conference on embedded computing (MECO-2018) – Including ECYPS-2018 (2018), pp. 1–5. https://doi.org/10.1109/MECO.2018.8406004
66. S. Beer, R. Ginosar, R. Dobkin, Y. Weizman, MTBF estimation in coherent clock domains. Proceedings of the international symposium on asynchronous circuits and systems (2013), pp. 166–173. https://doi.org/10.1109/ASYNC.2013.19
67. ASIP Designer (2021), https://www.synopsys.com/dw/doc.php/ds/cc/asip-brochure.pdf
68. T. Sato, S. Chivapreecha, P. Moungnoul, K. Higuchi, An FPGA architecture for ASIC-FPGA co-design to streamline. Process. IDSs. 412–417 (2017). https://doi.org/10.1109/cts.2016.0079
69. A.S. Hussein, H. Mostafa, ASIC-FPGA gap for a RISC-V core implementation for DNN applications. Proceedings of the 3rd novel intelligent and leading emerging sciences conference (NILES-2021) (2021), pp. 385–388. https://doi.org/10.1109/NILES53778.2021.9600503
70. The OpenCL specification. Khronos OpenCL working Group (2019). https://www.khronos.org/registry/OpenCL/specs/2.2/html/OpenCL_API.html
71. V. Mekkat, A. Holey, P.C. Yew, A. Zhai, Managing shared last-level cache in a heterogeneous multicore processor. Parallel architectures and compilation techniques – Conference proceedings (PACT) (2013), pp. 225–234. https://doi.org/10.1109/PACT.2013.6618819
72. S. Harutyunyan, T. Kaplanyan, A. Kirakosyan, A. Momjyan, Design and verification of autoconfigurable UART controller. Proceedings of the 2020 IEEE 40th international conference on electronics and nanotechnology (ELNANO-2020) (2020), pp. 347–350. https://doi.org/10.1109/ELNANO50318.2020.9088789
73. T.K. Kaplanyan, A novel pulse synchronizer design with the proposed sync cell model. Proc. RA NAS NPUA Ser. Tech. Sci. **74**(4), 464–470 (2021)
74. V.Sh. Melikyan, M. Martirosyan, A. Melikyan, G. Piliposyan. 14nm educational design kit: Capabilities, deployment and future. Proceedings of the 7th small systems simulation symposium 2018, Niš, Serbia, February 12–14 (2018), pp. 37–41
75. T.K. Kaplanyan, L.A. Mikaelyan, A.A. Petrosyan, A.M. Momjyan, et al, Design of video processing platform with interchangeable input-output interfaces. 2019 IEEE 39th international conference on electronics and nanotechnology: Proceedings (ELNANO-2019) (2019), pp. 201–205. https://doi.org/10.1109/ELNANO.2019.8783420

Chapter 6
Design of Digital Integrated Circuits by Improving the Characteristics of Digital Cells

6.1 General Issues in Design of Digital Integrated Circuits by Improving the Characteristics of Digital Cells

6.1.1 Importance of Design of Digital Integrated Circuits by Improving the Characteristics of Digital Cells

It is known [1] that standard cell (SC) libraries are currently widely used in IC design process. Any digital circuit (DC) to some extent includes digital ICs, such as simple logic operation; sequential, high-speed, power-saving, input/output (I/O) signal control; and other cells.

As a result of the wide use of SC libraries, in order to increase efficiency and reduce errors in the design process of their components, a large number of additional conditions are proposed. Standard cells repeatedly go through the stages of design, improvement, and optimization during the entire process of their creation (Fig. 6.1).

During the creation of SC library cells, in addition to basic design rules, such as design rule check (DRC) and layout versus schematic design (LVS), special attention is also paid to a number of other characteristics of SCs, such as:

- Compatibility of power supply and grounding cables with future designs
- Static and dynamic power consumption
- Routability
- Access to I/O cells
- Compatibility with other logical cells

In addition to the main libraries of SCs, which comply with the design rules and are optimized according to specifications, libraries with different heights (Fig. 6.2) and threshold voltages (Fig. 6.3) are also developed for the same technological process.

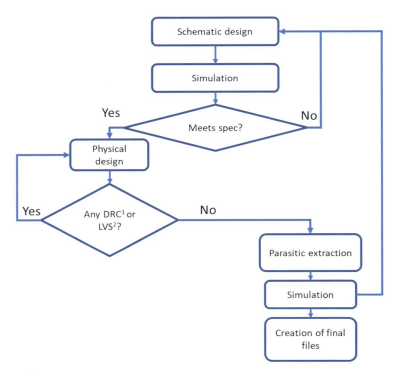

Fig. 6.1 SC design process

Fig. 6.2 SCs with different heights

1. Design rule check
2. Layout vs. schematic

The use of SCs with different threshold voltages and heights, dense I/O cells, different layout optimizations, and different power and supply structures in the same design leads to additional difficulties:

- Incompatibility of power supply network in the case of cells with different heights
- Errors arising from placement of cells
- Reducing routability of I/O cells

6.1 General Issues in Design of Digital Integrated Circuits by Improving...

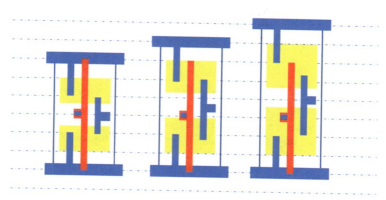

Fig. 6.3 Standard cells with different threshold voltages

- Large area of the design
- High power consumption of the design

The requirements for efficient IC design are:

Optimization of standard I/O cells: This is done in important steps by modifying the structure to obtain SCs with better characteristics in the result of further increasing the routability of the design and saving routing resources, as well as checking the I/O cells of already designed and optimized cells to avoid design problems for the creation of inter-cell distance rules (ICR) of the design.

Minimization of circuit area: The simultaneous use of SCs with different heights in one design, as experience shows [2, 3], significantly reduces the area of the circuit, because, mainly for reasons of performance, circuits use cells with a large height more often, which in turn leads to an increase of circuit area.

Power consumption optimization: Just as in the case of using cells with different heights, the use of lower cells in some places reduced the area of the circuit, so the use of high cells on critical paths leads to optimization of power consumption of the circuit, due to which, however, a decrease in performance may be observed.

Summing up, it should be noted that due to the widespread use of digital cells in the process of designing integrated circuits, optimization of their parameters is important. Therefore, the improvement and optimization of characteristics of digital cells, which are one of the important issues of the present, are the most important challenge for the effective design of ICs.

6.1.2 Current State and Issues of Design of Digital Integrated Circuits by Improving the Characteristics of Digital Cells

Effective development of ICs by improving the characteristics of digital cells is currently one of the current design problems. Improving the characteristics of digital cells includes various directions, the most popular of which are increasing the performance of cells, improving routability and accessibility of I/O cells, reducing dynamic and static power consumption, etc.

There are a number of well-known methods for effective development and optimization of ICs by improving SC characteristics [2–13], which can be divided into two main groups: pre-design [4–9] and post-design [2, 3, 10–13]. Among the most popular methods currently available are:

- Prediction and optimization of standard I/O cell accessibility using deep learning (DL) algorithms [4]
- Optimization of design results by using SCs with different heights in the same design [14]
- Due to the use of SCs with low power consumption, optimization of the power consumed in the design [15]
- Addition of extra metals by calculating timing parameters of the circuit [16]

Prediction and Optimization of Accessibility of Standard I/O Cells Using Deep Learning Algorithms

Currently, the large number of IC design rules and the complexity of the required checks lead to the demand for the use of ML algorithms in evaluation and optimization of the accessibility of standard I/O cells. In the case of different methods [17–19], using ML and DL techniques, SC I/O accessibility is estimated with high accuracy, and approaches to improve it are proposed, such as SC floorplan optimization, creation of placement distance rules, generation of routing files, etc.

One of the methods of evaluation and optimization of the accessibility of standard I/O cells of ML is the prediction of violation of design rules (DRV) based on their pattern [4, 20]. The peculiarity of this approach is that it is the first one that takes the pattern of I/O cells as a feature for the estimation of DRV.

The proposed approach can generally be divided into three main steps:

- Creation of data for the ML process
- Model training
- Detailed model-driven placement

Two ML models are considered: I/O pattern recognition (PR) and design-specific I/O pattern recognition (DSPR), of which I/O takes only I/O patterns as the only input data, and DSPR includes more design rules to increase prediction accuracy. Later, model-driven placement places the cells in such areas that no design problems arise [21]. Here, self-named training data is needed to perform supervised learning. Therefore, pre-designed designs and their DRC results are used in learning process (Fig. 6.4) [4].

Fig. 6.4 Learning and placement processes

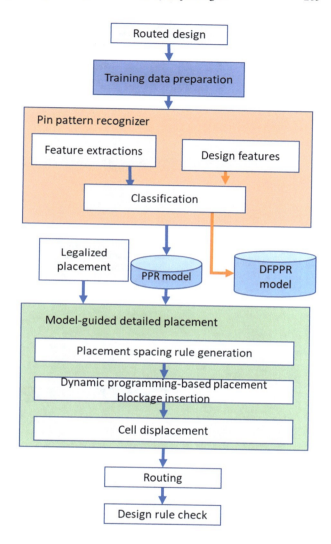

After generating the ML learning graph data, according to the method, a training data set is created from each DRC error place, centered on the short-circuit error section of the second metal layer (M2). In general, to apply the I/O pattern, a given section of the design is taken as m x n, where m is double of the unit height (UH) and n is equal to UH (Fig. 6.5) [4].

Then the received image is divided into pixels, the height and width of which are chosen according to the minimum distance of the first metal layer (M1), so that one pixel does not include two different M1 I/O cells [22, 23]. Then the value of each pixel is calculated, taking into account how much of it is occupied by the I/O cell

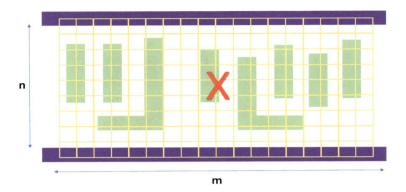

Fig. 6.5 A cell divided into pixels

Fig. 6.6 The area occupied by I/O cell within a pixel

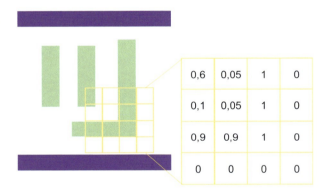

(Fig. 6.6) [4, 24], after which the processed information is transferred to the specifics generation module (SGM).

SGM represents a DL network consisting of two packet and two maximum unifying layers. The two packet layers separate the key specifics of I/O cell patterns, while the maximum unifying layers are used to reduce the number of specifics. After the specifics of I/O cells are extracted, they are smoothed and passed to the input of the classification algorithm. It represents a fully connected neural network and a sigma function.

The sigma function is considered the last layer of the classification algorithm, which scales the output value of the neural network between 0 and 1. Such scaling directs the model to understand whether a given I/O pattern will cause M2 short connection [25–28].

After the PR model is obtained, using DSPR, the presented approach helps the detailed placement tool to prevent receiving problematic pattern not to have short-connection circuit of M2. DSPR consists of three main steps:

- Creation of placement ICR
- Creation of routing constraints (RC) by dynamic programming
- Specification of cell movement

The ICRs generated during cell placement are described by the required number of distance locations (RNL) (6.1) [4, 29]:

$$\text{RNL}\left(A_i^m, A_j^n\right) = k : \qquad (6.1)$$

Equation (6.1) [4] describes the RNLs in the ICR between the i-th (A_i^m) cells with "m" direction and the j-th (A_i^m) cells with "n" direction. As a result of the method operation, the placement of cells is continuously changed in a stepwise manner until the probability of a short circuit predicted by the M2 model for the placed cells is less than 0.5 (Fig. 6.7) [4].

This method of evaluation and optimization of accessibility of standard I/O cells based on ML is fast and reliable, but has some disadvantages, namely:

- Application of a large number of sample designs for training ML model, since the effectiveness of the model directly depends on the amount of data used for training.
- Need to train the model in case of any change in cell floorplan and/or I/O cells because in this case the application of the previous model may lead to an incorrect result.
- RNL iterative addition of ICRs during design.
- Only M2 short-circuit fault prediction using PR and DSPR, because after having the pattern of standard I/O cells, it is possible to predict more design rule violation.
- In the process of creating a ML model, the training data, placed and routed designs, are entered as graphical data.

This results in necessity to perform the step of extracting the pattern of I/O cells from graphical data.

Taking into account the above, other methods aimed at improving the accessibility of cell pins were further developed, which, in addition to predicting and correcting errors at M2 metal level, also perform routability optimization at other metal levels, estimate the cost of the circuit, and improve IC characteristics by using SCs of different heights.

Optimization of Design Results by Using Standard Cells with Different Heights in the Same Design

It is known [14, 30, 31] that by using cells with different heights in the same design, certain parameters of the designed circuit are improved, namely:

- As a result of the use of cells with greater height, due to the increase in power consumption and the increase in area, it is possible to obtain a faster circuit.

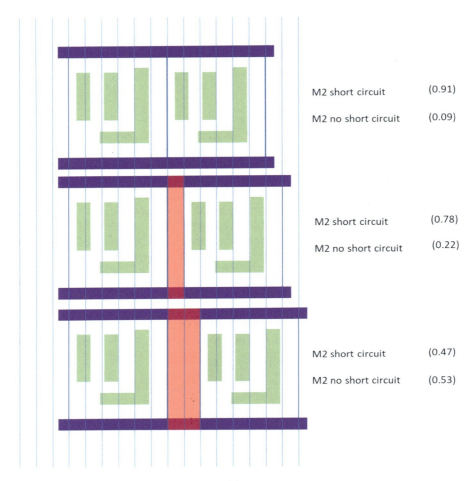

Fig. 6.7 Short circuit probabilities by cell position

- As a result of the use of cells with a smaller height, at the expense of deterioration of performance, it is possible to get a more dense and power-saving design (Fig. 6.8) [3].

In addition to deterioration of performance, cells with a small height have one more disadvantage. Such cells are more likely to have routability and I/O cells' accessibility issues because they include fewer routing paths due to their small size. Such cells can be drawn wider. In order to obtain a larger fan out, such an approach will increase the parasitic capacitances (PC) of gates and interconnects, which will lead to excessive power consumption and increased area [32].

Taking the mentioned problems into account, it can be said that the use of high and low SCs in the same design can have a positive effect on IC performance in

6.1 General Issues in Design of Digital Integrated Circuits by Improving... 287

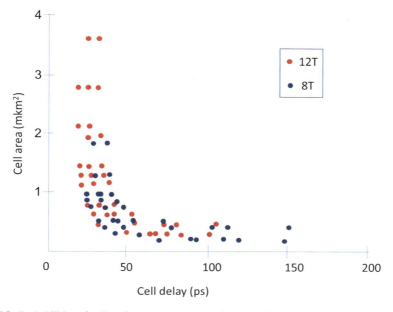

Fig. 6.8 Probabilities of cell performance and area ratio by position

places where it is more important, and in places that require denser placement, it is possible to use low cells.

There are various methods to improve the applicability of cells of different heights in the same design [14, 30, 31]. According to one such method [3], the main emphasis is placed especially on cells that differ from each other in height by not a full measure.

The application of this method starts by taking register transfer level (RTL) description of ready design, the timing parameters, logical and physical models of SC library, technological information including different heights of cells, area of placement and routing (P&R) part, and the aspect ratio of the design as input files. The final goal of the method's work is the orderly arrangement of logical cells on rows corresponding to their height. In addition to placement, the goal of the method is to obtain the smallest possible area and power consumption of the design under the conditions of the same performance [33–36].

In addition to the ones listed above, some other checks and conditions are also applied during the PR of the design to improve the performance of the method, namely:

- In order to ensure manufacturability, each fragment of SC placement with a certain height should have a minimum height of two units [37, 38].
- In order to ensure the continuity of nwell, each fragment must consist of an even number of rows [39].

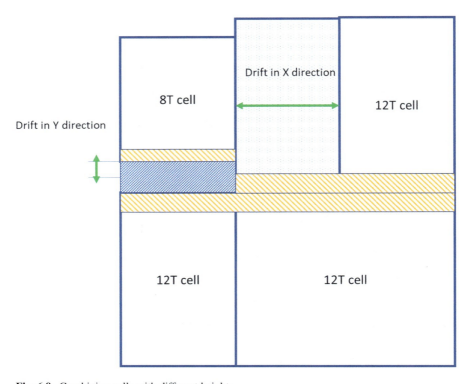

Fig. 6.9 Combining cells with different heights

- Each fragment must match the structure of metals and polysilicon layer (poly) in the design.
- The horizontal distance between two fragments should be at least 4 units.
- There should be enough distance in the vertical direction between two fragments so that the power and grounding buses of SCs of the first and second fragments do not touch each other (Fig. 6.9) [3]. In Fig. 6.9 [3], the minimum inter-metal distance (MID) of M2 is 64 nm, and the thicknesses of buses are 45 nm and 64 nm, respectively, in the case of low and high cells. In this case, even though the difference in the widths of buses is smaller than the MID, the minimum distance d must be 64 nm in order to fall on routing paths of metals.
- Divider cells should be placed to maintain minimum horizontal and vertical distances.

According to the method, at the beginning of the operation, the RTL description is taken, and the gate level description (GLD) design is synthesized using SC logic descriptions with different heights. Before starting the P&R of design process, the layout exchange format (LEF) file of physical description of cells is modified in order to break the closed loop of uncertainty between the floorplan and the height of the cells. The heights of all cells are changed to be equal to cells with the smallest

6.1 General Issues in Design of Digital Integrated Circuits by Improving...

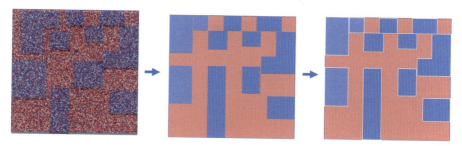

Fig. 6.10 Regular placement of cells of different heights

height, but the previous values are preserved. The modified LEF file is written to the computer's memory as mLEF (modified LEF). Such an approach eliminates uncertainties and allows using the current industrial P&R means for further processes according to the method [40]. As the initial SC logical descriptions are used during placement, P&R tools estimate and optimize the design with high accuracy. With this method, no conditions specific to the use of SCs with different heights are applied during primary placement, so that P&R tool freely selects the most suitable cells for area and timing parameters from libraries of SCs with different heights. Thus, the high cells are used in time-critical parts, and the low cells are used in area-critical parts.

After the first stage of placement, the placement area is divided according to SCs with a specific height, taking into account the area of dividing cells. In this process, the cell height values are restored to their original value [41–43]. Then, there are two ways to adjust the placement of cells:

- By cell movement
- By changing the height of the cell

The placement of the cells is considered regular if each of them is in the range of height, anticipated for it (Fig. 6.10) [3].

After the initial placement of cells is completed, the operation of the method can be divided into three stages:

- Division of floorplan and definition of areas
- Placement and regulation by calculating timing parameters
- Conversion from mLEF file to original LEF file

The method of optimizing the design results by using the presented standard cells with different heights in the same design is highly effective in placing and legalizing cells, but it also has drawbacks, such as:

- *Changing the cell description file with the LEF extension*, as a result of which the files are modified two times, and according to the method description, such a change should be made during the execution of each design.
- *Creation of separate placement sections for cells of each height*, as a result of which the physical design of a circuit is divided into parts in which cells of different heights are placed at a certain distance from each other.

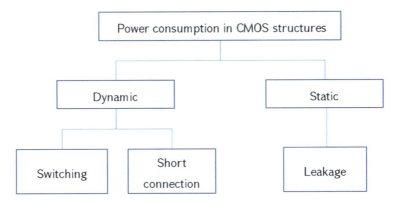

Fig. 6.11 Power consumption in CMOS circuits

- *After the initial routing and segmentation, the second routing is needed*, the reason being that the modified mLEF file is used first, then the detailed placement needs to use the original LEF file, too.

Due to these reasons, methods were later developed to ensure the applicability of SCs with different heights in the same design, while predicting the power consumption of the circuit and selecting optimal cells to obtain better results.

Standard Cell Library Characteristics Extraction with Optimal Power Consumption Using a Neural Network

As it is known [15, 44, 45], power consumption in circuits can be divided into two groups (Fig. 6.11) [19]—dynamic and static.

The power consumption in ICs is highly dependent on the power consumption of SCs, so to ensure low power consumption, cells need to be as optimal as possible.

Currently, there are various methods for reducing SC power consumption [20–24]. Some of them [15, 43] are based on the reduction of power consumption during the design of circuits, while others [44, 45] are based on ensuring their low consumption at the initial stage of SC design.

One of the SC power consumption reduction methods [15] is noteworthy, which by applying the DL technique to an already ready SC library, as a result, returns more suitable SC library and the most acceptable constraints to the given design.

In general, the library selection process for a design with optimal/near-optimal power consumption of the described method [15] can be divided into three main parts (Fig. 6.12) [15]:

- Training of power consumption estimation model
- Estimation of power consumption
- Library optimization

According to the method, presented in Fig. 6.12 [15], some input data are first read, namely, the different SC libraries, descriptions of several circuits, frequencies

6.1 General Issues in Design of Digital Integrated Circuits by Improving...

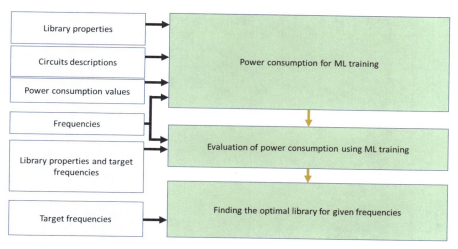

Fig. 6.12 Block diagram of method operation and input information

of circuits, and power consumption of each of circuits. The input information is then given to power cost estimation module, which includes a DL network. At the end of the training, in order to estimate the power consumption of any design, there is a need to provide the method with the SC library, the description of the new circuit, and the required frequency as input data. Information about the SC library here includes:

- Height of cells in them
- Heights of p and n domains
- Supply voltages

The *power consumption estimation model* includes a two-layer neural network, one hidden and one output layer (Fig. 6.13) [15]. Data input into a neural network is performed with n number of input features, and data output is performed with y output features. SC library data and circuit information are considered as input features for model training. The output of the model is the total power consumption of the circuit.

DL model is presented in the following way [15]:

$$f(x) = W_0 * h_1 + b_0, \qquad (6.2)$$

where $f(x)$ and h_1 are the results of output and hidden layers, respectively, W is the weight matrices, and b is the initial orientation values of layers. A neural network also contains a non-everywhere differential nonlinear continuous activation function [15]:

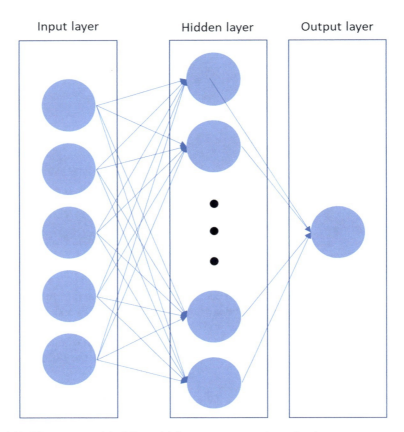

Fig. 6.13 The structure of the ML model for power consumption estimation

$$f(x) = \max(0, x). \tag{6.3}$$

The TensorFlow library [25] was used to build the power estimation model. The model uses the adaptive moment estimation (Adam) Stochastic optimization algorithm to reduce the number of errors between the actual and predicted values [26]. In order to exclude over-adaptation, the method of omitting some neurons during training is also used [27].

After getting power consumption estimation model, according to the method, the description of the most optimal/near-optimal library is searched among the imported libraries. All 24 imported libraries are used in parallel for this process and are given power estimation model as input data. According to the method, all libraries are checked in a simple step-by-step way, and as a result the selected library is returned (Fig. 6.14) [15].

6.1 General Issues in Design of Digital Integrated Circuits by Improving... 293

Description of optimal process

1. **begin**
2.   **for** for each library
3.   **while** Predicted frequency == Target frequency
4.   Power = Predicted power
5.   **if** Power < Power best value
6.   Registration of the given library
7.   Power best value= Power
8.   **end for loop**
9.   Return library parameters
10. **end**

Fig. 6.14 Sequence of library selection steps

Evaluating the Efficiency of Standard Cell Library Characteristics Extraction Method with Optimal Power Consumption Using a Neural Network

2020 training examples were used to evaluate the performance of SC library characteristics extraction method with optimal power consumption using a neural network. Each training data includes six columns: five input data and one output data.

The used libraries have the following parameters (Table 6.1) [15]:

- Heights of SC libraries expressed by the number of routing paths: 7, 9, 12, and 14
- The ratio of P and N domains in cells: 1, 2, and 3
- Supply voltages: 0.8 V and 1.2 V

After obtaining library parameters, nine conditional circuits were synthesized using them, the operating frequencies of which were selected from the range of 0.1–1 GHz. Then, training data was generated from the set of ISCAS-89 [46] test circuits, their timing parameters were calculated, and those designs, the difference of which from the required value was less than 0, were removed from the data list (Table 6.1) [15].

In order to adjust the weights and initial values of neural network, the input data were randomly divided in the ratio of 80:20 as training and testing data, respectively.

In order to evaluate the final efficiency of the model, the mean square error (MSE) (6.4) [15] and the determination coefficient (DC) (6.5) [15] are calculated:

Table 6.1 Properties of the applied SC libraries

Number of gates	Height	P/N value	Voltage (V)
1	7	1	1.2
2	7	1	0.8
3	7	2	1.2
4	7	2	0.8
5	7	3	1.2
6	7	3	0.8
7	9	1	1.2
8	9	1	0.8
9	9	2	1.2
10	9	2	0.8
11	9	3	1.2
12	9	2	0.8
13	12	1	1.2
14	12	1	0.8
15	12	2	1.2
16	12	2	0.8
17	12	3	1.2
18	12	3	0.8
19	14	1	1.2
20	14	1	0.8
21	14	2	1.2
22	14	2	0.8
23	14	3	1.2
24	14	3	0.8

$$\text{MSE} = \frac{1}{n} \sum_{i=1}^{n} (y_i - y'_i)^2, \tag{6.4}$$

$$\text{DC} = 1 - \frac{\sum_i (y_i - y'_i)^2}{\sum_i (y_i - y''_i)^2}, \tag{6.5}$$

where y'_i is the actual value, y'_i is the predicted value, y''_i is the average of actual values, and n is the number of all training data. According to the definition, the accuracy of the model is higher when the MSE has a minimum value and the DC has a maximum value [47–49] (Table 6.2).

As a result of model formation with ten different train-test data, the average value for MSE was 0.304, and 0.962 for DC.

Three new circuits from ISCAS-89 [29]—s526, s713, and s15850—were used for model validation. As a result of simulation, two different values of power consumption were obtained as a result of synthesis with target frequencies. In the end, the MSE values were calculated again, according to which:

6.1 General Issues in Design of Digital Integrated Circuits by Improving... 295

- Good results were obtained for s526 and s713, 0.227 and 0.096, respectively.
- A relatively large value of 1.230 was obtained for s15850.

As a result of evaluations, the following became clear.

With the library where the supply voltage is 0.8 V and the cell height is minimal, optimal designs are obtained in the frequency range of 100 kHz to 500 MHz. Although reducing the supply voltage can reduce power consumption, it also slows down the circuit, resulting in timing violations. Therefore, libraries with a high supply voltage (1.2 V) and a large cell height are suitable for designing high-frequency circuits, between 700 MHz and 1 GHz.

At the end of the operation of the described method, the generalized data are presented in the form of a table (Table 6.3) [15].

Table 6.2 ML training data generated from conventional designs

Circuit name	Number of inputs	Number of outputs	Number of DFFs	Number of gates	Function	Time (s)
S208	11	2	8	96	Digital fractional multipliers	27
S349	9	11	15	161	Traffic light controller	32
S832	18	199	5	287	Unknown	38
S1238	14	14	18	508	Combinational circuit	40
s1423	17	5	74	657	Unknown	67
S5378	35	49	179	2779	Unknown	53
s9234	36	39	211	5597	Real chip based	56
s13207	31	121	669	7951	Unknown	70
s35932	35	320	1728	16065	Unknown	361

Table 6.3 Data obtained as a result of the method

| Frequency (Hz) | s526 | | s713 | | s15850 | |
	Predicted optimal library	Real optimal library	Predicted optimal library	Real optimal library	Predicted optimal library	Real optimal library
100 K	2	2	2	2	2	2
500 K	2	2	2	2	2	2
1 M	2	2	2	2	2	2
5 M	2	2	2	2	2	2
10 M	2	2	2	2	2	2
15 M	2	2	2	2	2	2
50 M	2	2	2	2	2	2
100 M	2	2	2	2	2	2
500 M	2	2	2	2	2	2
700 M	2	1	2	1	2	1
1 G	2	1	2	1	2	3

With the method of extracting the characteristics of the library of standard cells with optimal power consumption using a neural network, it is possible to extract the optimal library for the given design with a high degree of accuracy. However, it has certain disadvantages, namely:

- To extract the optimal library information, a large number of libraries are read, which may have many parameters. Along with changing the parameters, it is necessary to change the ML model and retrain the whole system [50, 51].
- The method does not work with similar high accuracy in the case of different circuits of the same order.
- The method essentially does not optimize the circuit and/or cells in the case of optimization at a given frequency [52].
- Not only is the optimization/calculation of timing parameters not included in the optimization process, but also those libraries that violate one or another timing value are ignored during the calculation [16].

In order to eliminate the listed drawbacks, approaches to reduce IC power consumption by optimizing the characteristics of digital cells were further developed, which, in addition to reducing the total power consumption of circuits, also reduce the number of libraries needed for optimization process and optimize timing parameters of the circuit.

A Method of Adding Additional Metals by Calculating Timing Parameters of a Circuit

Currently, additional limitations are imposed during IC manufacturing due to the very small size of circuits being produced [53–58]. In addition to the fact that the manufactured circuit must have accurate DRC, LVS results, parasitic characteristics in the acceptable range, sufficient timing parameters, low power consumption, etc., it is important that it also meets special manufacturing rules. Among the methods aimed at increasing performance in designs are:

- Addition of cells without logic
- Increasing the size of interconnections where possible
- Addition of additional metals
- Verification and optimization of antenna effect
- Limitation of the minimum area of layers

Some of the listed methods do not have huge effect on the operation and timing parameters of a circuit, while others have quite an effect and can lead to deterioration of timing, power, and other parameters.

The addition of additional metals is one of the methods that has a great impact on timing parameters of a circuit. The expediency of using this method is conditioned by several circumstances, namely:

- Ensuring the manufacturability of the circuit for advanced technological processes
- Filling the missing gaps of the circuit with "dummy" metals

6.1 General Issues in Design of Digital Integrated Circuits by Improving...

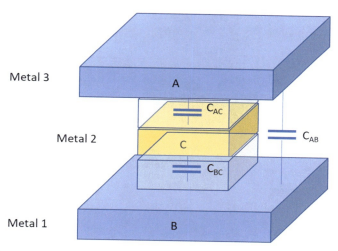

Fig. 6.15 Area capacitances

- Meeting the limitation of minimum density of metals in design
- Ensuring regular equalization of a circuit

Although the addition of additional metals is inevitable in current IC design and manufacturing processes, it leads to parasitic elements—capacitances that negatively affect circuit performance, power consumption, and timing parameters.

Metals added to a circuit can generally cause three types of parasitics: area, side, and edge.

Area capacitances arise when two conductors are located in different metal layers, and their projections on the ground surface overlap each other (Fig. 6.15) [53].

In Fig. 6.15, metal C creates parasitic capacitances with metals A and B, respectively, C_{AC} and C_{BC}, as well as in metals A and B itself, C_{AB}. In general case, the parasitic C_m capacitance, created by metals in metal layers l_1 and l_2 with cross-sectional area s, is calculated by (6.6) [53], where P_{12} is the area capacitance per unit area and is a function of s.

$$C_{12} = P_{12}(s) \times s. \tag{6.6}$$

Side capacitance can occur between two conductors that are in the same metal layer and have horizontal overlapping (Fig. 6.16) [53].

In Fig. 6.16 [53], C_1 and C_2 are, respectively, the capacitances created by metals A and B with metal C, and C_3 is the reciprocal capacitance of metals A and B. The side capacitance C_s is calculated by (6.7) [53], where l is the length of overlap of metals and $P_l(d)$ is the capacitance per unit length, depending on distance d between metals.

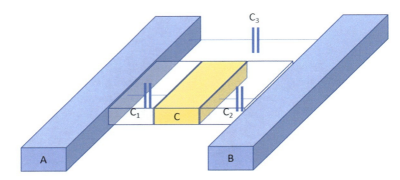

Fig. 6.16 Side capacitances

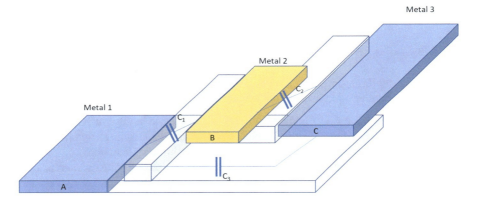

Fig. 6.17 Edge capacitances

$$C_s = P_{12}(s) \times s. \tag{6.7}$$

Edge capacitance can occur between conductors that are located on different metal layers, have overlap of parallel edges, but do not have overlap of area projections (Fig. 6.17) [53].

In Fig. 6.17, metal C together with metals A and B create edge capacitances C_1 and C_2, and A and B together create capacitance C_3. In general case, the capacitance C_e between layers l_1 and l_2 is calculated by (6.8) [53], where $P_{12}(d)$ and $P_{21}(d)$ are the edge capacitances per unit length depending on distance d between metals.

$$C_e = P_{12}(d) \times l + P_{21}(d) \times l. \tag{6.8}$$

The capacitance of a circuit to the ground point of all wires is called "equivalent" and can be calculated by analyzing the network using the conductance matrix.

Thus, the metals added to circuits for these purposes lead to additional parasitic characteristics and deterioration of timing parameters; therefore, when adding these

6.1 General Issues in Design of Digital Integrated Circuits by Improving... 299

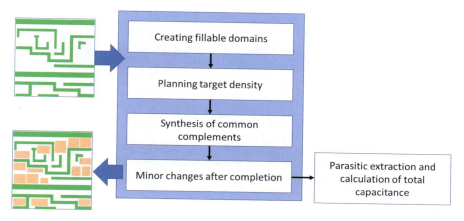

Fig. 6.18 Main operation of the method

metals, it is necessary to monitor timing parameters and optimize them to avoid problems [59].

Currently, there are many methods of adding additional metals with the reduction of parasitic characteristics, which reduce the parasitic capacitances and/or speed up the operation of the circuit by using certain algorithms [60–62].

One of such methods is the introduction of additional metals by calculating timing parameters (TCMI) [53]. In this method, the input information is a graphical description of an already placed and routed design, on which it introduces additional metals, reducing the equivalent capacitance of the circuit in both extreme and remaining wires.

The main operation of the described method can be divided into four processes (Fig. 6.18) [53]:

- Creation of fillable domains
- Target density planning
- Synthesis of common additions
- Minor changes after addition

The method of getting parasitic extraction embedded in the method contributes to the rapid extraction of the equivalent capacitance during the creation of complementary domain and planning processes of target density.

Creation of filling domains is the first operating process of the method. Since the input physical descriptions contain only locations of wires, it is necessary to separate the polygons to be filled corresponding to input wires given from such a design. The domains to be filled are those in which the addition of metals does not cause violations of the minimum distance of metals compared to the wires already in the circuit. At the beginning of the process of creating domains, the dimensions of the given wires are widened by the method in the amount of the minimum distance of metals, and then the coordinates of polygons limiting the remaining free spaces are taken. The resulting polygons generally have a rather complex structure with many

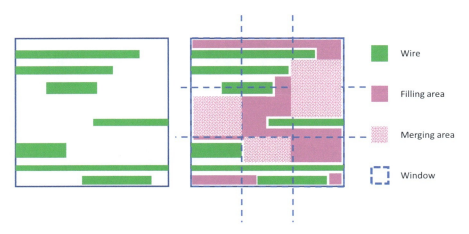

Fig. 6.19 Uniting domains to be filled

corners and sometimes holes inside. Because of this, it is necessary to perform one more step: converting polygons to rectangles [31]. It should be noted that the quality of subsequent addition of additional metals is highly dependent on the underlying conversion, as a result of which the improved method of converting polygons to rectangles (IPRC) is used during this process. It iteratively considers each given vertex and efficiently creates rectangles [31].

Later, the IPRC method transforms into direction-specific polygons-to-rectangles (DPRC) method, where the images are described as "wide" and "high" rectangles depending on the preferred routing direction of a given metal layer, in the horizontal and vertical directions, respectively. Then, on the basis of the obtained results, the vertical and horizontal revision algorithm is applied, which unites the domains to be filled, which have a common side with each other (Fig. 6.19) [53].

The performed verifications have shown that, compared to IPRC method, DPRC significantly increases the upper limit of metal density and allows introducing better solutions in the future, which is important especially in the case of introducing additional metals in lower metal layers.

The next step in the described method is to plan the target density. In general, for efficient calculation of minimum metal density, the selected verification window is taken with dimensions w x w and shifted to area of the circuit by $\frac{w}{2}$ step. To satisfy such a verification, it is advisable to divide the circuit by $\frac{w}{2} \times \frac{w}{2}$ when adding the additions and immediately apply the minimum metal density condition to obtained windows. However, on the other hand, larger window sizes are more efficient for synthesis of additions, in terms of providing greater flexibility in adding additions and splitting local additions. Therefore, in the described method, instead of using the above subwindows, a new approach is described: target density planning, which creates a window-specific density value in each window, taking into account two conditions:

6.1 General Issues in Design of Digital Integrated Circuits by Improving...

- Reduction of capacitances with decisive wires
- Reduction of the overall capacitance of the circuit

Assume a W window system consisting of m × n W_{ij} subwindows is given, where $1 \leq i \leq m$ and $1 \leq i \leq n$. In this case, (6.9) [53] will work for planning the target density.

$$\min \sum_{i,j} O_{i,j} D_{i,j} - \min\left\{D_{i,j}^{\max} - D_{i,j}\right\} \qquad (6.9)$$

where for $W_{i,j}$ window:

- $D_{i,j}^{\text{wire}}$ is the density of wires before adding the filling.
- $D_{i,j}^{\max}$ is the maximum density value.
- $D_{i,j}$ is the target density value.
- $D_{\min}$ is the minimum density value.
- $O_{i,j}$ is the criticality factor of $W_{i,j}$.

$$D_{i,j}^{\text{wire}} \leq D_{i,j} < D_{i,j}^{\max} \qquad (6.10)$$

where:

- a_{ij}^c is the critical wire area inside $W_{i,j}$ window.
- a_{ij}^{nc} is the area of the noncritical wires inside $W_{i,j}$ window.
- w^c and w^{nc} are a_{ij}^c and a_{ij}^{nc} coefficients, respectively.
- e is an infinite small constant.

Equation (6.9) [53] consists of two parts (6.11 and 6.12) [53], where the first part contributes to reducing the target density in deterministic and/or multiwired windows.

$$D_{i,j} + D_{i,j+1} + D_{i+1,j} + D_{i+1,j+1} \geq 4 * D_{\min}, \qquad (6.11)$$

$$O_{i,j} = \begin{cases} e, & a_{ij}^c = 0, a_{ij}^{nc} = 0 \\ w^c * a_{ij}^c + w^{nc} * a_{ij}^{nc} \end{cases} \qquad (6.12)$$

Actually w^c should be smaller than w^{nc} to ensure a lower filling density in windows with high criticality. In addition, e provides larger densities in the neighborhood of critical windows. Part 2 of (6.12) [53] contributes to maintaining the minimum difference between the maximum density value and the planned density value. Such planning of target densities creates initial values for all windows in a more optimal and flexible way, but, as can be seen from the formulas, they are not usually linear, so in their solution, it is advisable to add certain additional parameters to ensure linearity (6.13–6.16) [53].

$$\min \sum_{i,j} O_{i,j} D_{i,j} - M, \qquad (6.13)$$

$$M \leq D_{i,j}^{\max} - D_{i,j}, \qquad (6.14)$$

$$D_{i,j}^{\text{wire}} \leq D_{i,j} < D_{i,j}^{\max}, \qquad (6.15)$$

$$D_{i,j} + D_{i,j+1} + D_{i+1,j} + D_{i+1,j+1} \geq 4 * D_{\min}. \qquad (6.16)$$

After creating the target densities, synthesis of metal additions is performed (i.e., additional metal layers that do not contain logical connections are added to free places of the circuit). The additions created at this stage are of particular importance in preventing the deterioration of timing parameters of the circuit, because they are responsible for the emergence of additional capacitances in the circuit. The most important points to be calculated in the process of creating additions can be considered [32–34]:

- Prevention of occurrence of areal capacitances between decisive wires and implemented additions
- Increasing possible metal distance between additions and other conductors
- Prevention of inserting additions in the middle of two parallel conductors
- Reduction of parallel linear length between additions and parallel conductors

After creating the initial values of variables in the algorithm for creating general additions, according to the difference between the windows and the maximum possible densities, the windows are classified in descending order. Then, the operation of iterative addition is performed on them (Fig. 6.20) [53].

Algorithm 1: Creation of additions by metal layers

Login. L metal layers W windows

1. Initialization of layer laws and verification of S distances
2. Classification of **W windows according to density difference**
3. **For** window w W **to do**
4.   while S >= minSpace **do**
5.   success <- windowFiller(w,S,D_{target})
6.   **if** success **then** break;
7.   S = S – steplength;
8.   Detect violations of w, return D if necessary
9.   Change the layer laws to include the results of the additions in the w window
10. Perform checks of laws and densities for the metal layer L
11. Return the results of additions to the L metal layer

Fig. 6.20 Iterative filling algorithm

After completion of additions, the algorithm of detailed review of additions also causes a considerable improvement, in which the algorithm for calculating the Euclidean distance between the domains to be filled and the decisive wires performs the main function.

When an optimal addition of fillings is performed in the circuit in the result of synthesis of common additions, it is still possible to have certain additions that are not so optimally arranged and create large reciprocal capacitance with decisive wires. In order to solve such problems, small local changes with computation of timing parameters are made in the described method. This process can be divided into two parts:

- With the calculation of timing parameters: transfer of additions
- With the calculation of timing parameters: displacement of additions

During the transfer of additions by calculating timing parameters, those additions that have a great effect on timing parameters are moved to domains with less effect. Here, the effect of filling in the given range on decisive wires and their equivalent capacitance is calculated. In the transfer process, $|A_i - A_{i1}|$ difference is defined where A_i is the original area of additions and A_{i1} is the one obtained as a result of the change. If the specified difference is greater than the specified minimum value, then the change is considered acceptable; otherwise, the previous value is maintained.

In the case of displacement of additions by calculating timing parameters, the method tries to find a more suitable location for each addition, reducing the decisive capacitance. Here, a one-way movement is made either in the direction of x-axis or in the direction of y-axis.

The presented method creates and places the metal fillers with high precision and also optimizes the equivalent capacitances between them and the critical wires. However, this method also has certain disadvantages, namely:

- No attention is paid to routing of the SC signals, but only decisive wires are considered. However, SC signals are also of great importance in the circuit design process, and the lack of specific requirements for them can lead to certain losses.
- In the case of creating additions, the voltage drop of the circuit is not calculated, which, in general, depends on the equivalent capacitance of the circuit, and as a result of improving the capacitance, it can also be improved.

Thus, a review of existing methods known from the literature shows that although they optimize SC libraries with a high degree of accuracy and generate optimal libraries for doing various designs, they still have limited coverage and/or do not optimize SCs sufficiently.

In particular,

- IC I/O accessibility checking algorithms partially/incompletely fix SC library cells, as a result of which later, during IC design, cases of violation of design rules derived from SCs increase.
- Methods of using cells with different heights in the same design to reduce power consumption make a large number of changes in the SC description files with the

condition of later post-modification and create new ML models for each design, resulting in an increase in design time and the number of description files. In addition, when choosing the optimal library from the point of view of power consumption for a given circuit, a large number of libraries are used, which leads to the extension of the time of the tools, and the human factor in the design process increases.

- Methods of creating metal fillings do not take into account the possible voltage drop during their creation and are not aimed at improving it, while their addition to the design can significantly improve the results of IC voltage drop and reduce the resources spent on it later.

For the reasons listed, with the improvement of the characteristics of digital cells, there is a need to develop new principles and means of effective IC design, which, taking into account the issues presented above, will allow to optimize circuits.

6.1.3 Proposed Principles for Efficient Design of Integrated Circuits by Improving the Characteristics of Digital Cells

Thus, in order to solve the aforementioned problems, the following principles are proposed for the effective design of integrated circuits by improving the characteristics of digital cells:

Unlike the methods known from the literature, which optimize several layers, optimize all metal layers. Since the methods known from the literature optimize only some metal layers, not considering the other layers, it is proposed to create an environment for checking and optimizing the accessibility of I/O pins.

The latter will check the I/O pins by creating special matching cases for cells and mark those cells that have accessibility problems. The main difference of this approach, compared to existing methods, is that all cells are collected in one complete design and combined according to a special pattern (Fig. 6.21), as a result of which an opportunity is created to check the routability of cells in dense placement conditions in cases of different directions of cells.

Using the ML algorithm, possible problems will be predicted as a result of cell routing, and inter-cell distance rules necessary for their solution will be created. In this case, the main difference compared to existing means is that the multi-class classification method is used during the training of ML model, which makes it possible to check and optimize other metal layers as well (Fig. 6.22) [35].

The use of filler cells created by ML during the placement of cells with different heights. This method, in contrast to the methods known from the literature, in the case of placement of cells with different heights, uses filler cells created by ML.

6.1 General Issues in Design of Digital Integrated Circuits by Improving... 305

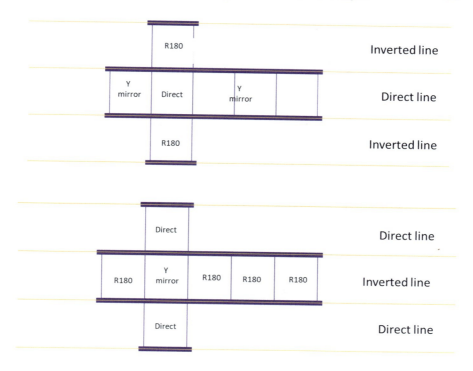

Fig. 6.21 Layout of cells in the verification process

Fig. 6.22 The ML model of multi-class classification

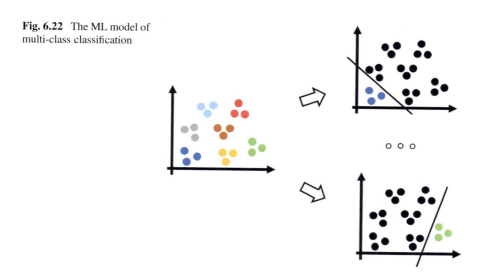

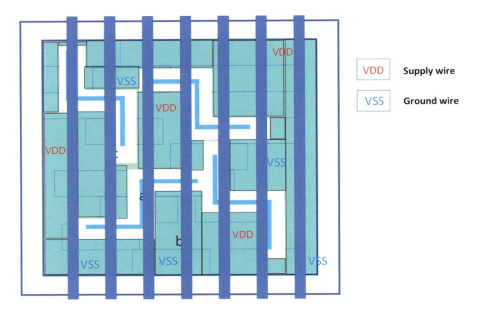

Fig. 6.23 Connection of metal fillers to power and grounding wires

As a result of this, it is possible to place the cells, without a big change in total IC area without violating the rules of power and grounding buses and main layers (Fig. 6.23).

The addition of metal fittings connected to the power and grounding grid for IC design process. Since metal additions increase the overall capacitance of the circuit, but have an important effect on other stages of IC design, it is possible to optimize the voltage drop on the power and grounding of the circuit when adding them. The main difference of this method from the others is that the metal additions added to the circuit are connected to the power and ground wires (Fig. 6.24), as a result of which the influence of rapidly changing signal wires in the design on the other wires is reduced, improving the voltage drop. Such a change significantly reduces the voltage drop at the expense of a certain deterioration of timing parameters.

By improving the characteristics of digital cells, the proposed principles of effective design of integrated circuits will allow to significantly improve the main technical characteristics and parameters of SCs: power consumption, designability, and parasitic characteristics.

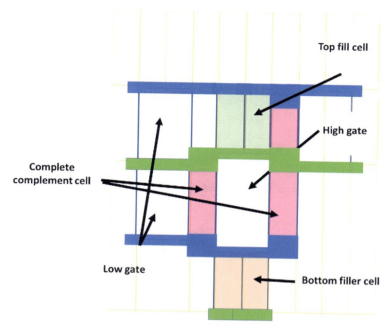

Fig. 6.24 Standard and special cell placement

Conclusions

1. Development of effective integrated circuit design tools by improving the characteristics of digital cells is relevant, as it can significantly improve the main parameters and characteristics of digital integrated circuit designs based on them.
2. Analysis of the existing approaches and means of developing means of effective design of integrated circuits by improving the characteristics of digital cells shows that in the latter, the important circuit parameters, designed based on them, power consumption, voltage drop in the power supply, routability, etc., are not sufficiently optimized. However, from the point of view of practical requirements of the design of modern digital integrated circuits, this degree of optimization is not enough, and especially from the point of view of efficiency, there is a need to develop new methods, algorithms, and software tools that are significantly superior to the existing ones.
3. By improving the characteristics of digital cells, the principles of effective design of integrated circuits have been proposed, which, at the expense of machine time costs and nonsignificant deterioration of timing parameters and area of the designed integrated circuit, allow to significantly improve the main parameters of integrated circuits designed based on them: power consumption, voltage drop in power supply, routability, etc.

6.2 Methods of Design of Digital Integrated Circuits by Improving the Characteristics of Digital Cells

As mentioned in Sect. 6.1, the accessibility of standard I/O pins plays an extremely important role in IC design process. Its improvement leads to easier circuit routing and reduced DRC violations. Two methods are used to improve the accessibility of standard I/O cells: relatively simple and predictive and optimizing with the use of ML algorithms.

6.2.1 Method of Optimizing the Accessibility of Standard I/O Cells

Optimization of I/O cell accessibility to SCs with an experimental design [63] is a relatively simple approach to check a ready SC library and mark cells with poor I/O access. This approach is used to cover the I/O accessibility of cells in a simple design, as well as to achieve performance compared to the I/O verification method [64]. Another advantage of this verification is that at the end of the design, it is possible to perform a single DRC and LVS on one final design file.

The operation of the method starts by reading logical and physical descriptions of relevant libraries, a simple RTL description, and the technology file (Fig. 6.25).

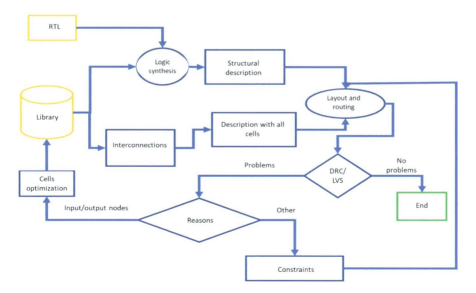

Fig. 6.25 The block diagram of the method of optimizing access to I/O cells SCs with an experimental design

6.2 Methods of Design of Digital Integrated Circuits by Improving... 309

```
CELLNAME_1 inst_CELLNAME_1_cntr ( .A(In[43]),
.B(In[63]), .C(Out[178]) );
 CELLNAME_2 inst_CELLNAME_2_cntr ( .A(In[40]),
.B(In[47]), .C(In[50]), .D(Out[188]));
 CELLNAME_3 inst_CELLNAME_3_cntr ( .A(In[41]),
.B(Out[198]));
 CELLNAME_4 inst_CELLNAME_4_cntr ( .A(In[48]),
.B(Out[32]));
 CELLNAME_5 inst_CELLNAME_5_cntr ( .A(In[60]),
.B(Out[78]));
 CELLNAME_6 inst_CELLNAME_6_cntr ( .A(In[100]),
.B(Out[19]), .C(Out[55]), .D(Out[44]),);
 CELLNAME_7 inst_CELLNAME_7_cntr ( .A(In[15]),
.B(Out[18]));
```

Fig. 6.26 An excerpt from the generated interconnect file

The purpose of reading a simple RTL description is to make it possible to connect the generated second circuit to it and implement the final LVS.

One of the most important parts of the method's operation is the creation of interconnections. At this stage, all cells of all libraries read are taken, five different instances of the same cell are created, and their I/O cells are randomly connected to each other. At the end, the corresponding symbols "center," "right," "left," "top," and "bottom" are added to the resulting five copies of the same cell, which will guide the tool to make the correct placement in later stages.

The final interconnect file contains the cell name, symbol, and connection order (Fig. 6.26).

Depending on the number of cells in the library, the circuit obtained during the creation of interconnect file may include between 5 and 10 thousand cells (Fig. 6.27).

In the next stage, placement is done keeping the signs of the cells, after which the cells are placed in groups on the entire area (Fig. 6.28).

After the regular placement of cells, the design of the file according to their interconnects is performed. At the end of this process, a rather dense routing of cell I/O cells is ensured (Fig. 6.29), which makes it possible to easily identify problematic cells and I/O cells in conditions of larger routing paths.

After the cells are designed, production-oriented changes (ORCs) are performed on the project, and final files are written, on which DRC and LVS are performed. As a result, cells with a large number of DRC violations are marked as problematic cells from the point of view of placement and routing [65].

Thus, using the method of optimizing accessibility of standard I/O cells with an experimental design, it is possible to identify possible problems arising during their routing through special combinations of cells, at the same time, without limiting the number of metal layers passing through the verification process [66].

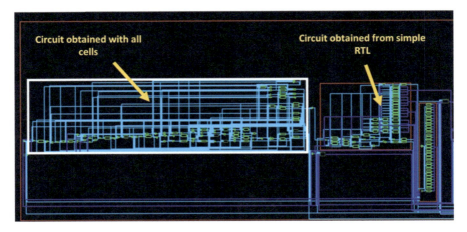

Fig. 6.27 Part of the synthesized circuit

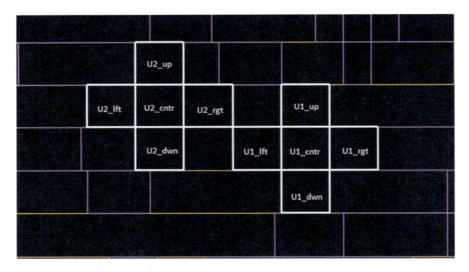

Fig. 6.28 Placement example of a circuit

Evaluation of the Effectiveness of the Method of Optimizing Accessibility of Standard I/O Cells with an Experimental Design

In order to evaluate the effectiveness of the proposed method for optimizing accessibility of standard I/O cells with an experimental design, the SAED 14 nm educational design kit with a gate width of 14 nm and containing no intellectual property was used [67]. At the initial stage of performance evaluation, the library was divided into three smaller parts, which included 50, 100, and 200 cells, respectively. While adding cells to the new libraries, the rules of Boolean algebra and digital design are respected for the purpose of synthesizing any RTL description with

6.2 Methods of Design of Digital Integrated Circuits by Improving... 311

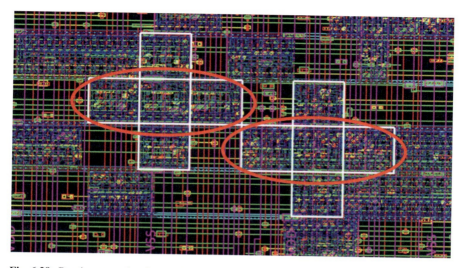

Fig. 6.29 Routing example of a circuit

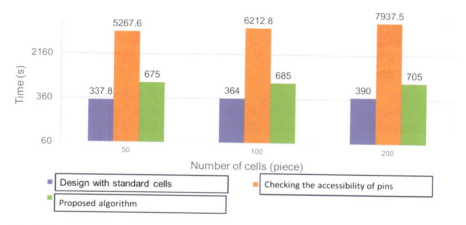

Fig. 6.30 Comparison of verification process times of existing and proposed methods

the resulting libraries. Then, with the new libraries obtained, three types of designs with simple RTL description, I/O accessibility verification algorithm, and application of the presented method were made. DRC and LVS checks were performed on the circuits obtained at the end of the design.

According to the experimental results, as a result of using the proposed method, the cell library verification speed increased by approximately 5784.3 s or approximately 9.4 times compared to the I/O cell verification method (Fig. 6.30). Under the conditions of the proposed placement of cells, the number of cases of their matching

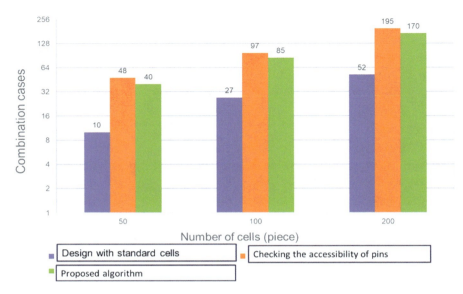

Fig. 6.31 Comparison of cases of cell matching of existing and proposed methods

increased by approximately 67 or approximately 59.2% compared to the experimental design with a simple RTL description (Fig. 6.31).

Thus, as a result of the application of the proposed method for optimizing accessibility of SC I/O cells in experimental design, it is possible to obtain an average speedup of approximately 9.4 times for cell verification compared to I/O verification algorithm and approximately 59.2% more matching cases compared to a design with a simple RTL description. It also makes it possible to check all metal layers instead of a single metal layer.

6.2.2 Enhanced Method for I/O Cell Accessibility Prediction and Design Optimization with Proposed Machine Learning

In order to predict the accessibility of I/O pins of SCs in the mid-design phase and to increase their routability during design, an enhanced method of I/O cell accessibility prediction and design optimization with machine learning is proposed. The operation of this method can generally be divided into three parts (Fig. 6.32). They are:

- Extraction of I/O coordinates from designs
- ML model training
- Creation of horizontal and vertical ICRs for SCs

6.2 Methods of Design of Digital Integrated Circuits by Improving...

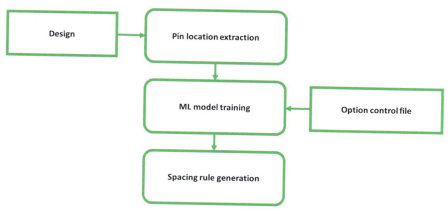

Fig. 6.32 Steps of an enhanced method for predicting the accessibility of ML I/O cells and optimizing the design

Table 6.4 The format for saving the obtained results

Row	Information/data			
	Source name	I/O cell name	Metal layer	Coordinates
1	HVT_ANDX1	A	2	{33.73 21.46} {33.73 22.27} {34.63 22.27} {34.63 21.46}
2	LVT_INVX8	X	1	{13.93 30.37} {13.93 31.18} {14.53 31.18} {14.53 30.37}
3	LVT_FDPSQX2	D	2	{34.93 37.66} {34.93 38.47} {35.83 38.47} {35.83 37.66}

Extraction of I/O coordinates from designs is one of the first and most important steps of the proposed method, because the accuracy of later-created ML model depends on the mentioned coordinates. Therefore, it is necessary that, on the one hand, the extracted coordinates are accurate, and on the other hand, they contain quite simple information, taking into account the speed of further processes. To extract I/O coordinates from design, the method first reads various designs in the memory, along with their corresponding DRC results. Then, using the simple commands of placement and routing tools, it extracts the reference name of the used cells [68], the metal layer creating I/O cells, and their coordinates. The obtained results are saved in a predetermined special format (Table 6.4). As during the extraction of I/O pins, not the topological description of design but the design library [69] and simple commands of placement and routing tool are used, this process does not take much time to perform.

After extracting the I/O coordinates, the proposed method creates a ML model to predict the problematic cells. As a result of model operation, it is possible to identify at most seven types of DRC violations:

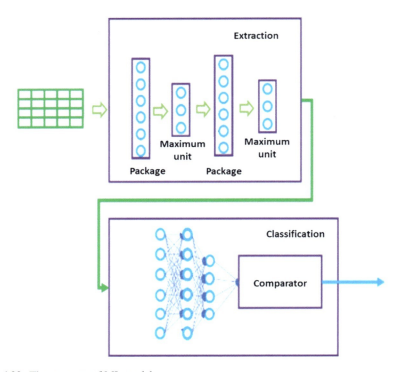

Fig. 6.33 The structure of ML model

- Distance at the end of M1 line
- Minimum distance of M1
- M1 short connection
- M2 short connection
- "Minimum distance of the same wire" of M2
- Minimum area of M3

In order to optimize the training time of the model, each of the listed violations can be turned off individually, deactivating the prediction made on it.

The ML model consists of two nodes: extraction of characteristics and classification (Fig. 6.33).

Four neural layers are used for characteristics extraction. The first one represents the packet layer, followed by maximum merging, second packet, and second maximum merging layers (6.17):

$$A_{i,j} = \max(D_{k,m}), k = [0, i], m = [0, j]. \qquad (6.17)$$

Both packet layers have training weights, the values of which change depending on the presence of DRC violations in the given segment. In the next step, the extracted characteristics are passed to the classification node. This node multi-

6.2 Methods of Design of Digital Integrated Circuits by Improving...

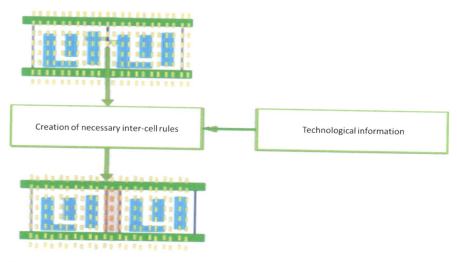

Fig. 6.34 ICR creation process

classifies the input data according to several labels. A one-versus-all type of multi-layer classification is used here [70, 71]. To apply the one-versus-all type of classification, the entire data set is divided into eight separate groups, with the training group labeled as "1" and all others as "0." In this case, when the data is entered into the classification algorithm, it goes through the process of binary classification, and in the case of a predicted violation, a "positive" value is generated, and in the case of all others, a negative value (6.18).

$$\text{Predicted value} = \begin{cases} > 0, \text{predicted violation} \\ < 0, \text{other violations} \end{cases}. \quad (6.18)$$

When the cells that have the probability of DRC occurrence during routing are separated from the design, the proposed method then creates ICRs for them to prevent the given violation. Later, these ICRs can be used in any design made with these cells. To prevent DRC violations, the method calculates the existing distance of I/O cells of problematic cells and then, reading the technology information, adds an ICR value such that the probability of this violation will be less than 0.5. For example, for cells U1 and U2 in Fig. 6.34, the method predicted a short circuit of M1, after which, to avoid the problem, it created an ICR in the amount of 2 units.

When the cells that have the probability of DRC occurrence during routing are separated from designs, the proposed method then creates ICRs for them to prevent the violation.

Later, these ICRs can be used in any design made with these cells.

During ICR creation, cell directions are considered to avoid excessive constraints. As a result, the process of creating ICRs can be represented by (6.19):

$$\text{ICR}(C_{1,x}, C_{2,y}) = A \tag{6.19}$$

where $C_{1,x}$ represents the first cell in the X direction and $C_{2,y}$ is the second cell in the Y direction. The necessary distance between them is defined by A number.

Thus, as a result of the application of the ML model, it is possible to predict and later also avoid routing problems arising during a specific arrangement, as well as to have a larger number of metal layers to be checked.

Evaluation of the Effectiveness of the Proposed Machine Learning Enhanced I/O Cell Accessibility Prediction and Design Optimization Method

SAED14nm [67] libraries were used to evaluate the effectiveness of the above method. With them, designs were carried out, in which the density of cells varies in the range of 30–40%. In this technological process, the number of cells for each threshold voltage is approximately ~1000. The number of ML data taken from designs was approximately 6000, which was then divided into training and testing lists with a ratio of 70:30. The system on which checks were made has the following parameters: 2 cores with a frequency of 2.4 GHz, Linux operating system. Then the obtained results were compared with simple design methods presented in [63] (Table 6.5).

From the point of view of metal layers different from M2, the simple design approach and the presented method were also compared (Table 6.5).

The accuracy of the ML model for all DRC types was also calculated using (6.20) (Table 6.6).

$$\text{accuracy} = \frac{\text{TP} + \text{TN}}{\text{TP} + \text{TN} + \text{FP} + \text{FN}} \times 100\%. \tag{6.20}$$

The values given in (6.15) have the following physical meaning:

- TP—number of correctly predicted DRC errors
- TN—number of correctly predicted error-free DRCs
- FN—number of incorrectly predicted DRC errors
- FP—number of incorrectly predicted error-free DRCs

Thus, according to experiments, by applying the proposed machine learning enhanced I/O cell accessibility prediction and design optimization method, it is possible to predict seven types of DRC violations in different metal layers. The application of the method allows to reduce the existing DRC violations in the design by approximately 47%, but as a result of the application of ML, the tool processing time increases by approximately 23% (Table 6.7).

6.2 Methods of Design of Digital Integrated Circuits by Improving... 317

Table 6.5 Comparison of results

Design	Common design approach			Approach presented in [2]				Proposed approach			
	DRC violations	M2 short circuit	Wire length (um)	DRC violations	M2 short circuit	Wire length (um)	During tool process (s)	DRC violations	M2 short circuit	Wire length (um)	During tool process (s)
D1	1971	61	5,532,549	1007	14	5,606,690	391	934	23	5,259,218	485
D2	572	7	5,518,484	232	0	5,586,467	385	267	4	5,241,467	472
D3	663	10	5,510,817	295	4	5,578,543	387	253	7	5,233,484	480
D4	796	19	5,503,626	288	5	5,573,211	388	312	18	5,230,246	475

Table 6.6 ML model accuracy

Check	Accuracy
Distance at the end of the M1 line	51%
M1 minimum distance	42%
M1 short circuit	61%
M2 short circuit	64%
M2 minimum distance of the same connection	57%
M3 minimum surface	49%
Cases without violations	68%

Table 6.7 Comparison of metal layers different from M2

	Common design approach			Proposed approach			
Design	DRC violations	M1–M3 violations	Wire length (um)	DRC violations	M1–M3 violations	Wire length (um)	During tool process
D1	1971	61	5532549	934	568	5259218	485
D2	572	7	5518484	267	103	5241467	472
D3	663	10	5510817	253	176	5233484	480
D4	796	19	5503626	312	446	5230246	475

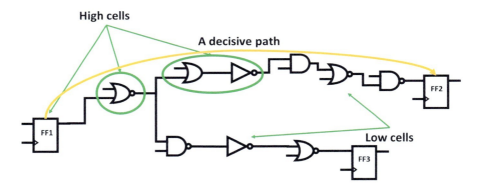

Fig. 6.35 The result of logical synthesis with cells of different heights

6.2.3 Optimization Method of Standard Cells for Digital Integrated Circuit Designs with Different Cell Heights

As already mentioned above, the use of cells with different heights in the same design increases the circuit performance and reduces its power consumption. Logical synthesis tools are easily able to choose the height cell during synthesis process with the greatest possible accuracy, which provides maximum performance and minimum power consumption on the given time path (Fig. 6.35). The main problem of such synthesis of circuits is that later the physical placement and routing tool does not have the capability to perform the necessary synthesis with such ease.

6.2 Methods of Design of Digital Integrated Circuits by Improving... 319

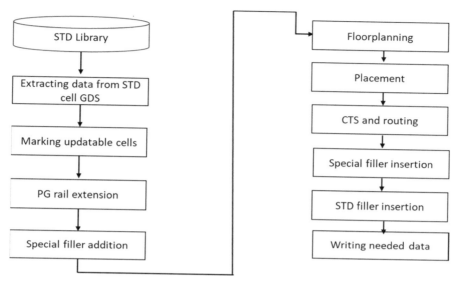

Fig. 6.36 The flow of optimizing SCs when doing design of cells having different heights

A method for increasing the accessibility of cells with different heights in the same design is proposed [72], which, by modifying the GDS file of physical description of cells in the given library, makes it possible to place them regardless of their height. The flow of the proposed method (Fig. 6.36) mainly consists of two parts: changing the power and grounding buses of SCs and implementing the design with modified cells [73–75].

Basically, SCs have the same structure of power and grounding buses [76]. Such a structure allows for easy connection of cells to circuit-level power and grounding grid during the design phase when placing the cells next to each other. And when combining cells in vertical direction, they are placed straight and upside down, sharing the given power or grounding bus [77–79].

When SCs of different heights are placed next to each other, the arrangement of their power and ground buses, as well as many other main layers, does not match correctly, which makes such placement impossible (Fig. 6.37).

As described, the combination of cells with different heights is unacceptable from the point of view of cell placement approaches and power and grounding buses [80–82]. The proposed approach to solve these problems, using ML algorithms, modifies the power and grounding grid of SCs as well as creates special filler cells for correct placement of cells with different heights.

The operation of the method begins with reading GDS description of SC library and separating layers used for placement and routing. The collected information is then processed into ML algorithm for reading and storage.

In the next stage, the ML model, reading the previously prepared information, specifies the metals belonging to power and grounding grid [83]. In this process, the

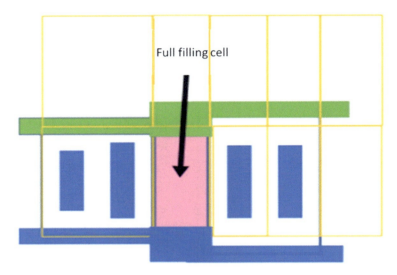

Fig. 6.37 Complications arising when combining cells of different heights

basic patterns of metal layers are also extracted: minimum distance, minimum angle length, minimum distance of the same wire, etc. Then, on those cells, where it is possible to add additional power and grounding metals, according to the method, following the rules, they are added. In general, it is possible to change the power and ground buses on cells in cases of which:

- There are no signal and/or power and grounding wires in the specified areas.
- The addition of power and grounding wires does not lead to violation of rules of metal layers.
- All minimum distance rules are kept.
- The additional supply and grounding buses are of the same type as the cell's buses.
- Added layers intersect with M1 and M2 routing paths.

All the cells, whose structure and the view of power and grounding buses do not correspond to the mentioned points, are left out of the change process [84, 85].

Another important circumstance for the change of SCs is the selection of their placement limit. Current placement and routing tools determine the legal location of cells when placing them, depending on their starting point [86, 87]. For this reason, in order to be able to place the modified cell using placement and routing tools, their starting point is raised and aligned with the starting point of the small height cells (6.21).

$$U_{x,y} = \{0 + Hx - Lx, 0 + Hy - Ly\}. \tag{6.21}$$

6.2 Methods of Design of Digital Integrated Circuits by Improving...

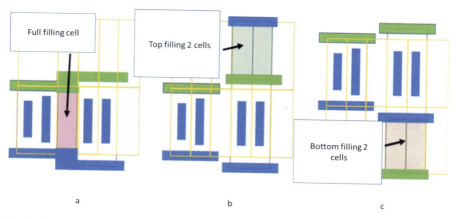

Fig. 6.38 Additional cells to be created

After the corresponding changes in the power and grounding grids, three types of filling cells are created (Fig. 6.38) for filling the lines in places of matching. These filling cells have two main uses:

- Keep one minimum distance between high and low cells
- Fill the free lines after placement and routing processes

If the free area is greater than one minimum distance, the filling cells can be combined.

The files created as a result of method operation are of three types: GDS descriptions of modified filling cells and added filling cells and the list of cells that have not been modified.

In order to design with the obtained library, first, using the initial logical description of cells, a logical synthesis is performed. As the logic synthesis tool tries to reduce circuit area and increase performance, it will select the required cell from the given libraries.

In the physical design phase, the minimum size of the placement grid is chosen according to the cells with the lowest height. After that, the circuit-level power and grounding buses are created. They are mostly M2 or higher metal layers. Then placement and routing is done, and then filling of empty places using the SC library and filling cells, created according to the method.

Thus, by using the proposed method, it is possible to apply them in the same design as a result of changing the physical descriptions of SCs with different heights. During the operation of this method, only GDS descriptions of SC library are modified, and then, using standard tools, placement and routing are performed.

Evaluating the Effectiveness of Optimization Method of SCs for the Implementation of Design with Cells of Different Heights

To evaluate the effectiveness of the proposed method, three designs were performed with libraries of 9T (low) and 12T (high) heights. At the end of the implementation

Table 6.8 Comparison of parameters of circuits with cells of different heights

Design	9T library	12T library	9T and 12T libraries together
Number of cells	2079586	2069610	2081234
Area (um^2)	386554.0594	486154.0594	399407.0619
Setup violations (ns)	−23.471	−4.473	−1.257
Cell density	0.73	0.64	0.67
Worst setup violations (ps)	−195	−83	−62
Wire length (um)	724438	698103	718639
Power consumption (nW)	2.61e+10	3.16e+10	2.96e+10
Leakage power (nW)	8.11e+06	11.26e+06	9.12e+06
During the process (m)	348.1	327.8	438.4

of designs, the physical and timing parameters of obtained circuits were compared (Table 6.8).

Thus, as a result of using the proposed method, it is possible to obtain approximately 19% optimization in circuit area and approximately 14.3% optimization in timing parameters, compared to circuits designed with only 12T and only 8T libraries, respectively. However, as a result of the changes made, the time spent for placement and routing increased by approximately 27%, and the power consumption increased by only 13%.

6.2.4 "Sleep Mode" Integration Method of Cells Using Neural Network for Low-Power Designs

As mentioned earlier, IC power consumption has two component parts: dynamic and static. One of the options used to reduce power consumption [14] is to disconnect SCs that are not currently used in the circuit from the power supply and grounding grid. Thus, the given part of the circuit will have as low power consumption as possible (there will be a certain amount of consumption due to leakage currents). However, implementing this approach requires either special libraries or more complex design steps.

In order to overcome the described challenges, the method of integration of "sleep mode" in SCs using a neural network is proposed [88]. The operation of this method (Fig. 6.39) can be divided into five main parts:

- Creation of format of input files and reading
- Extraction of technology-specific parameters with the DL algorithm
- Creation of special cells
- Indication of cells to be changed
- Changing cells and extracting final files

6.2 Methods of Design of Digital Integrated Circuits by Improving... 323

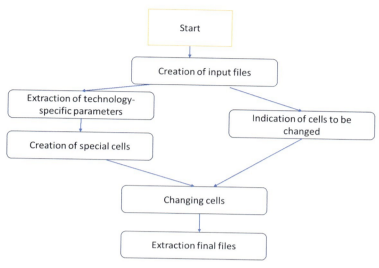

Fig. 6.39 Operating principle of "sleep mode" integration method in SBs using a neural network

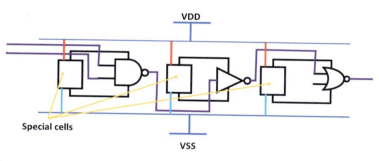

Fig. 6.40 Circuit of cell connection

According to this method, all cells that can be optimized are disconnected from the main power source. Their connection is provided by special cells. That is, conditional "false" supply and grounding buses occur, the cells of which are connected (Fig. 6.40).

In the first step, the proposed method reads the GDS description of the SC library and extracts the coordinates of metals, interconnects, and blocking layers from it [89]. The output information of this stage is the positions of different layers and, if any, the names of I/O cells attached to them. The information is then given to a data extraction algorithm, which includes a three-layer DL model. The first of the layers is the input layer, followed by two hidden layers and one output layer [90]. The output layer of the model returns the min values extracted for the following metal rules: distance, overlap, edge distance, area, and U-distance. These data are later used in the process of creating special types of cells [91, 92].

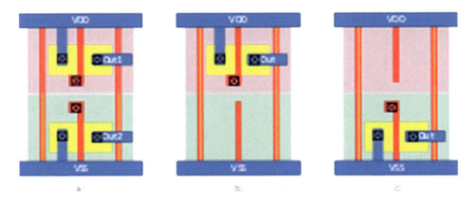

Fig. 6.41 Control cell views

The creation of special cells is the next stage of the proposed method. At this stage, the method, using the extracted values, creates special power and grounding control cells, which can be of three types (Fig. 6.41):

- P-MOS and N-MOS control
- P-MOS control, N-MOS filling
- P-MOS filling, N-MOS control

In the next step, the method reads the cells of the SC library one by one and marks those that have the characteristics of the change. They are:

- SC has a power and/or ground connection close to the edge.
- No large amount of routing barriers and metals near cell edges.

For the cells that meet the mentioned requirements, the power and grounding grid is changed, and special cells are added [93]. If the power and grounding connections of the cell are located on one side of it, then only one special cell is added; otherwise, according to the appearance of the connections, two cells are added (Fig. 6.42). After that, the SC initial power and ground connections are disconnected and connected to the control cell.

At the end, the GDS description containing all the cells of the library is written, including the modified ones.

Thus, the proposed method changes the appearance of the initial SCs of the library through the DL network and integrates the "sleep mode" control into them. In the future, the use of such cells in designs will provide an opportunity to manage the connection of cells to power and grounding grid, thus reducing static and dynamic power consumption.

Evaluating the Effectiveness of Cell "Sleep Mode" Integration Method Using Neural Network for Low-Power Designs

To evaluate the effectiveness of the method, designs with two libraries of different heights were implemented. One of them is 12T high; the other is 8T. In the completed designs, the cells were used in one case without the proposed change,

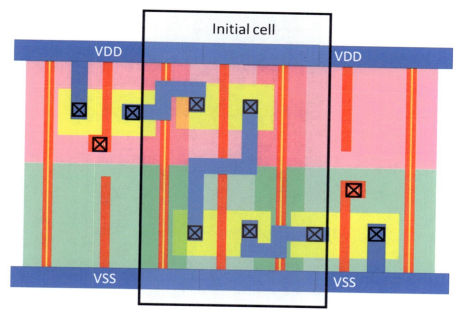

Fig. 6.42 The final appearance of SB after the addition of control cells

in the other case, including it. According to the results (Table 6.9), high-height cells have greater capabilities for change, but the optimization in power consumption is approximately 5–6% greater for low-height cells. According to the obtained results, the total power consumption can be reduced by approximately 12%, but due to the added special cells, the area of circuits increases by approximately 5–28%, and the timing parameters deteriorate by approximately 16% on average.

Thus, as a result of using the proposed method, it is possible to obtain an average power consumption saving of approximately 12% at the expense of approximately 5–28% increase in area and approximately 16% deterioration of timing parameters.

6.2.5 Method of Adding Metal Fillers for Digital IC Design Flow

As already mentioned in Sect. 6.1, the addition of extra metals in the circuit is extremely important for increasing its productivity. However, this process causes an increase in the overall parasitic capacitance of the circuit, which affects its performance and speed. On the other hand, during circuit operation, it is important to have a small value of voltage drop across its buses, because, otherwise, the connected voltage from high-level power and ground buses will drop greatly before reaching the SCs [94].

Table 6.9 Results of designs with modified and unmodified cells

	Design 1				Design 2			
	8T library		12T library		8T library		12T library	
	Using the presented method	Without using the presented method	Using the presented method	Without using the presented method	Using the presented method	Without using the presented method	Using the presented method	Without using the presented method
Total power consumption (nW)	3.91e+10	3.39e+10	5.21e+10	5.73e+10	2.94e+10	2.47e+10	3.75e+10	4.38e+10
Leakage power (nW)	5.91e+06	4.8e+06	6.29e+06	7.1e+06	4.18e+06	3.50e+06	6.58e+06	5.49e+06
Total violation of calibration (ns)	−36.19	−27.18	−23.97	−26.83	−26.83	−19.85	−16.74	−14.57
Worst calibration violation (ps)	−157	−129	−110	−114	−114	−89	−78	−74
Area (um^2)	536987.183	467178.846	869635.04	402740.387	402740.387	341160.559	643549.559	519411.74

According to those listed, it is important to be able to add extra metals to the circuit while paying attention to the voltage drop on its buses [95].

For the implementation of the mentioned function, the method of adding metal additions in IC design process is proposed [96, 97], by which it is possible to add metals to the already completed circuit, considering and optimizing the voltage drop.

The steps of the proposed approach can be broadly divided into four main groups:

- Separation of domains to be filled.
- Detection of overlaps of domains of power and grounding wires to be increased.
- Reducing the voltage drop, increase metals and interconnects.
- Calculating the target density, add additions.

For the proposed method, in the first stage, the graphic description of the already placed and routed circuit is the input data, and the coordinates of the domains, where the minimum distance violations will not occur as a result of adding metal fillers, are the output data. To obtain this result, first, a routing blocking layer is drawn on the entire area of the circuit, then the layer is cut with the minimum distance accuracy in the areas where there are routed metals (Fig. 6.43). The routing barriers obtained in this process have a rather complex structure. They consist of different pieces and have a non-standard appearance. For this reason, they go through another process and are divided into "long" and "high" quadrangles [53]. The coordinates of the resulting quadrilaterals are saved as domains to be filled, and the routing constraints are deleted [97, 98].

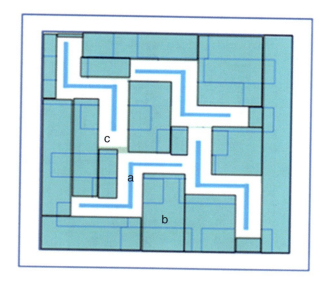

Fig. 6.43 Cutting routing barriers

When the coordinates of the areas to be filled are already known, the next step is to identify the power and/or grounding wires located near them or crossing them. It is important for this process that the rectangle to be filled in:

- The area meets the minimum requirements of the given metal layer.
- The width and length should be greater than the sum of dimensions of minimum thickness and distance for the given metal layer.

All fillable quadrilaterals that meet the specified conditions are marked as possible power and grounding domains.

In the next step, already marked domains are given a special attribute—"owner." This indicates which wire the specified domain should be connected to later: power or ground. In order to connect the domain metals to any wire, it is important that:

- It is close to the given wire or has overlap.
- Its addition does not lead to maximum density problems.

Since one power or ground wire may be adjacent to or overlap multiple filler metals, a wire-to-delta distance is defined between the wire and the center of gravity of the filler domain. The filler domains are then classified by delta distance. After that, the method physically connects the power and ground wires to metals of the fill domain, according to the already mentioned "owner" characteristic. To maintain the total density, the method also records the current metal density in memory [99, 100]. At the end, additions are added to the remaining free domains to maintain the minimum density rules by calculating timing conditions [53]. At the end of the design, the output information is the physical description of the design containing metal fillings (Fig. 6.44).

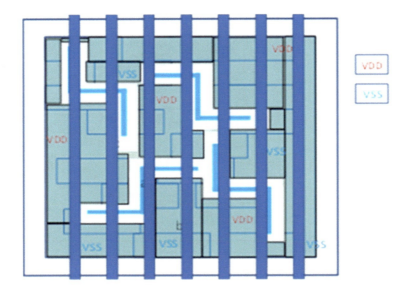

Fig. 6.44 The final view of the design

Thus, as a result of using the proposed method, it is possible to add additional metals to the already placed and routed circuit. Added metals are attached to the power or ground wires, as appropriate. This makes it possible to reduce the voltage drop in the circuit due to the creation of a protective effect on the wires.

Evaluating the Effectiveness of Adding Metal Filler Method for IC Design Flow
To evaluate the effectiveness of the proposed approach, five types of designs were implemented, using three different methods: standard, presented in [53], and using the proposed method. All experiments were performed on a Linux machine with 4 cores each at 3.4 GHz. The general comparative results are given in Table 6.10. A comparison of voltage drop values of the circuit was also made for standard and recommended approaches (Table 6.11).

Thus, using the method of adding metal fillers for IC design flow, it is possible to obtain an approximate 11.9% reduction in the voltage drop on the power and ground wires of the circuit. However, as a result of added metal additions, the overall circuit capacitance increases by approximately 4.4%, and the tool process time increases by 27.8% as a result of additional processes.

Conclusions
1. A method of improving access to I/O standard cells with an experimental design was proposed, in which case, due to special logical synthesis, placement, and routing using all cells in the library, a reduction of verification time by approximately 9.4 times was obtained, at the expense of only 11.2% reduction of matching cases.
2. An enhanced machine learning I/O cell accessibility prediction and design optimization method was proposed, in which, due to consideration of inter-cell distance constraints, the number of design rule violations decreased by approximately 47% at the expense of approximately 23% increase in machine time.
3. A method of optimizing standard cells for designs with elements of different heights has been developed, in which, due to their simultaneous application, an average optimization of approximately 14.3% of timing parameters of the circuit has been obtained, at the expense of only approximately 12.8% loss in power consumption.
4. A "sleep mode" integration method based on low-power design and the use of neural networks was proposed, in which, due to the introduction of a controlled power network of cells, an average reduction of approximately 12% of power consumption was provided at the expense of approximately 5–28% increase in IC area.
5. A method of adding metal fillers to the design of integrated circuits was developed, in which, due to the connection of special filler layers to power and ground buses, an average voltage drop reduction of approximately 11.9% was obtained at the expense of an approximately 4.4% increase in the parasitic capacitance of the circuit.

Table 6.10 Comparative results of the methods

Design	Simple design			Method design [53]			Proposed method design		
	Wire length (um)	Total capacitance (pF)	Time (m)	Wire length (um)	Total capacitance (pF)	Time (m)	Wire length (um)	Total capacitance (pF)	Time (m)
Design 1	311,250	10712.9	7.9	305,667	10883.66	8.8	312,373	11258.1	12.3
Design 2	742,837	34627.1	30.28	750,166	39523.68	31.44	758,960	41256.75	37.4
Design 3	62,739	1499.7	1.54	64,903	1558.17	1.59	64,749	1648..24	2.27
Design 4	150,507	2960.71	3.17	149,464	2969.20	3.55	154,783	3109.64	4.61
Design 5	268,315	5514.83	6.21	275,425	5705.39	6.97	247,614	5908.41	7.49

Table 6.11 Comparison of voltage drop values

Design	Simple design		Proposed method design	
	Wire length (um)	Voltage drop %	Wire length (um)	Voltage drop %
Design 1	311,250	1.127	312,373	1.004
Design 2	742,837	1.432	758,960	1.103
Design 3	62,739	0.974	64,749	0.914
Design 4	150,507	0.993	154,783	0.861
Design 5	268,315	1.014	247,614	0.947

References

1. L.M. Naga, P. Mullangi, Design and development of an ASIC standard cell library using 90nm technology node. 2018 international conference on computer communication and informatics (ICCCI) (2018), pp. 1–6. https://doi.org/10.1109/ICCCI.2018.8441222
2. S. Hougardy, M. Neuwohner, U. Schorr, A fast optimal double row legalization algorithm. International symposium on physical design (ISPD '21) (2021), pp. 1–5
3. K. Darav, N.A. Kennings, D. Westwick, L. Behjat, High performance global placement and legalization accounting for fence regions. 2015 IEEE/ACM international conference on computer-aided design (ICCAD) (2015), pp. 514–519
4. T.-C. Yu et al., Pin accessibility prediction and optimization with deep-learning-based pin pattern recognition. IEEE Trans. Comput.-Aided Des. Integr. Circuits Syst. **40**(11), 2345–2356 (2021). https://doi.org/10.1109/TCAD.2020.3040078
5. S. Fang, C. Tai, R. Lin, On benchmarking pin access for nanotechnology standard cells. 2017 IEEE computer society annual symposium on VLSI (ISVLSI) (2017), pp. 237–242. https://doi.org/10.1109/ISVLSI.2017.49
6. M. Danigno et al., Proposal and evaluation of pin access algorithms for detailed routing. 2019 26th IEEE international conference on electronics, circuits and systems (ICECS) (2019), pp. 602–605. https://doi.org/10.1109/ICECS46596.2019.8965194
7. V.B. Suresh, P.V. Kumar, S. Kundu, On lithography aware metal-fill insertion. Thirteenth international symposium on quality electronic design (ISQED) (2012), pp. 200–207. https://doi.org/10.1109/ISQED.2012.6187495
8. X. Bai et al., Timing-aware fill insertions with design-rule and density constraints. IEEE Trans. Comput.-Aided Des. Integr. Circuits Syst., 1–4 (2021). https://doi.org/10.1109/TCAD.2021.3133854
9. Y. Chen, H. Jiao, Standard cell optimization for ultra-low-voltage digital circuits. 2019 international conference on IC design and technology (ICICDT) (2019), pp. 1–4
10. W.-T. Wong, K. Singh, J. Huisken, J.P. de Gyvez, Power and variation improved near-Vt standard cell library for 28-nm FDSOI. 2019 IEEE SOI-3D-subthreshold microelectronics technology unified conference (S3S) (2019), pp. 1–2. https://doi.org/10.1109/S3S46989.2019.9320687
11. N. Mamikonyan, N. Melikyan, R. Musayelyan, IR drop estimation and optimization on DRAM memory using machine learning algorithms. 2020 IEEE East-West design & test symposium (EWDTS) (Varna, 2020), pp. 1–4. https://doi.org/10.1109/EWDTS50664.2020.9224772
12. X.X. Huang, H.C. Chen, S.W. Wang et al., Dynamic IR-drop ECO optimization by cell movement with current waveform taggering and machine learning guidance. 2020 IEEE/ACM international conference on computer aided design (ICCAD) (San Diego, 2020), pp. 1–9
13. Y. Lin, B. Yu, D.Z. Pan, High performance dummy fill insertion with coupling and uniformity constraints. IEEE TCAD **36**(9), 1532–1544 (2017). https://doi.org/10.1145/2744769.2744850

14. S.A. Dobre, A.B. Kahng, J. Li, Design implementation with noninteger multiple-height cells for improved design quality in advanced nodes. IEEE Trans. Comput.-Aided Des. Integr. Circuits Syst. **37**(4), 855–868 (2018)
15. S.H. Lim et al., Generating power-optimal standard cell library specification using neural network technique. 2017 IEEE Asia Pacific conference on postgraduate research in microelectronics and electronics (Prime Asia) (2017), pp. 101–104. https://doi.org/10.1109/PRIMEASIA.2017.8280374
16. Y.C. Liu, C.Y. Han, S.Y. Lin, J.C. Li, PSN-aware circuit test timing prediction using machine learning. IET Comput. Digit. Techniq. **11**(2), 60–67 (2017)
17. Q. Zhou, X. Wang, Z. Qi, Z. Chen et al., An accurate detailed routing routability prediction model in placement. Proceedings of the Asia symposium on quality electronic design (ASQED) (2015), pp. 119–122
18. P. Debacker, K. Han, A.B. Kahng, H. Lee et al., Vertical M1 routing-aware detailed placement for congestion and wirelength reduction in sub-10nm nodes. Proceedings of ACM/IEEE Design Automation Conference DAC) (2017), pp. 1–6
19. F. Tabrizi, N.K. Darav, S. Xu, L. Rakai et al., A machine learning framework to identify detailed routing short violations from a placed netlist. Proceedings of ACM/IEEE Design Automation Conference (DAC) (2018), pp. 1–6
20. Z. Qi, Y. Cai, Q. Zhou, Accurate prediction of detailed routing congestion using supervised data learning. Proceedings of 2014 IEEE international conference on computer design (ICCD) (2014), pp. 97–103
21. F. Tabrizi, N.K. Darav, L. Rakai et al., Detailed routing violation prediction during placement using machine learning. Proceedings of international symposium on VLSI design, automation and test (VLSIDAT) (2017), pp. 1–4
22. W.T.J. Chan, Y. Du, A.B. Kahng et al., BEOL stack-aware routability prediction from placement using data mining techniques. Proceedings of IEEE international conference on computer design (ICCD) (2016), pp. 41–48
23. Z. Xie, Y.H. Huang, G.C. Fang, H. Ren et al., RouteNet: Routability prediction for mixed-size designs using convolutional neural network. Proceedings of IEEE/ACM international conference on computer-aided design (ICCAD) (2018), pp. 1–8
24. W.T.J. Chan, P.-H. Ho, A.B. Kahng, P. Saxena, Routability optimization for industrial designs at sub-14nm process nodes using machine learning. Proceedings of ACM international symposium on physical design (ISPD) (2017), pp. 1–5
25. J. Seo, J. Jung, S. Kim, Y. Shin, Pin accessibility-driven cell layout redesign and placement optimization. Proceedings of ACM/IEEE design automation conference (DAC) (2017), pp. 1–6
26. M.M. Ozdal, Detailed-routing algorithms for dense pin clusters in integrated circuits. IEEE Trans. Comput.-Aided Des. Integr. Circuits Syst. **28**(3), 340–349 (2019)
27. W. Ye, B. Yu, D.Z. Pan, Y.C. Ban et al., Standard cell layout regularity and pin access optimization considering middle-of-line. Proceedings of Great Lakes symposium on VLSI (GLSVLSI) (2015), pp. 1–5
28. Y. Ding, C. Chu, W.K. Mak, Pin accessibility-driven detailed placement refinement. Proceedings of ACM international symposium on physical design (ISPD) (2017), pp. 1–4
29. X. Xu, B. Cline, G. Yeric, B. Yu, et al., Self-aligned double patterning aware pin access and standard cell layout co-optimization. IEEE Trans. Comput.-Aided Des. Integr. Circuits Syst. **34**(5), 699–712 (2015)
30. S. Dobre, A.B. Kahng, J. Li, Mixed cell-height implementation for improved design quality in advanced nodes. Proceedings of ICCAD (Austin, 2015), pp. 854–860
31. L. Guo, Y. Cai, Q. Zhou, X. Hong, Logic and layout aware voltage island generation for low power design. Proceedings of ASP DAC (Yokohama, 2007), pp. 666–671
32. J.A. Ellis, Embedding rectangular grids into square grids. IEEE Trans. Comput. **40**(1), 46–52 (1991)

33. B. Kahng, S. Kang, H. Lee et al., High performance gate sizing with a signoff timer. Proceedings of ICCAD (San Jose, 2013), pp. 450–457
34. K. Han, A.B. Kahng, J. Lee et al., A global-local optimization framework for simultaneous multi-mode multi-corner clock skew variation reduction. Proceedings of DAC (San Francisco, 2015), pp. 1–6
35. R.L.S. Ching, E.F.Y. Young, K.C.K. Leung, C. Chu, Postplacement voltage island generation. Proceedings of ICCAD (San Jose, 2006), pp. 641–646
36. H. Wu, M.D.F. Wong, I.-M. Liu, Timing-constrained and voltage island-aware voltage assignment. Proceedings of DAC (San Francisco, 2006), pp. 429–432
37. J. Alpert, A. Devgan, C. Kashyap, A two moment RC delay metric for performance optimization. Proceedings of ISPD (San Diego, 2000), pp. 73–78
38. V. Kashyap, C.J. Alpert, F. Liu, A. Devgan, PERI: A technique for extending delay and slew metrics to ramp inputs. Proceedings of TAU (Monterey, 2002), pp. 57–62
39. Y. Lin et al., MrDP: Multiple-row detailed placement of heterogeneous sized cells for advanced nodes. Proceedings of ICCAD (Austin, 2016), pp. 1–8
40. OpenCores. [Online]. Available: http://opencores.org. Accessed 11 Aug 2014
41. H. Wu, M.D.F. Wong, Improving voltage assignment by outlier detection and incremental placement. Proceedings of DAC (San Diego, 2007), pp. 459–464
42. H. Wu, I.-M. Liu, M.D.F. Wong, Y. Wang, Post-placement voltage island generation under performance requirement. Proceedings of ICCAD (San Jose, 2005), pp. 309–316
43. F. Beeftink, P. Kudva, D. Kung, L. Stok, Gate-size selection for standard cell libraries. 1998 IEEE/ACM international conference on computer-aided design. Digest of technical papers (IEEE Cat. No.98CB36287) (San Jose, 1998), pp. 545–550
44. F. Ye, F. Firouzi, Y. Yang, K. Chakrabarty, et al., On-chip droop-induced circuit delay prediction based on support-vector machines. IEEE Trans. Comput.-Aided Des. Integr. Circuits Syst. **35**(4), 665–678 (2016)
45. J. Zhou, S. Jayapal, B. Busze, L. Huang, et al., A 40 nm dual- width standard cell library for near/sub-threshold operation. IEEE Trans. Circuits Syst. I Regular Pap. **59**(11), 2569–2577 (2012)
46. S. Kajihara, K. Kinoshita, I. Pomeranz, R. Sudhakar, Combinationally irredundant ISCAS-89 benchmark circuits. Circuits and systems, 1996. ISCAS '96., Connecting the world., 1996 IEEE (1996) (Vol. 4), pp. 632–634. https://doi.org/10.1109/ISCAS.1996.542103
47. M. Anis, M. Allam, M. Elmasry, Impact of technology scaling on CMOS logic styles. IEEE Trans. Circuits Syst. II Analog Digit. Sign. Process **49**(8), 577–588 (2002)
48. Kingma, J. Ba, Adam: A method for stochastic optimization. CoRR (2014) (Vol. abs/1412.6980), pp. 1–4
49. S. Kung, R. Puri, Optimal P/N width ratio selection for standard cell libraries. 1999 IEEE/ACM international conference on computer aided design. Digest of TEChnical Papers (Cat. No.99CH37051) (San Jose, 1999), pp. 178–184
50. M. Abadi et al., Tensorflow: Large-scale machine learning on heterogeneous distributed systems. CoRR (2016) (Vol. abs/1603.04467), pp. 1–4
51. P. Bastani, K. Killpack, L.C. Wang, E. Chiprout, Speedpath prediction based on learning from a small set of example. 2008 45th ACM/IEEE design automation conference (Anaheim, 2008), pp. 217–222
52. A.A. Yarygin, Current issues of machine learning with the support of intellectual agents in decision-making tasks. Automat. Prob. Ideas Solut. C, 1–62 (2017) (in Russian)
53. B. Jiang et al., FIT: Fill insertion considering timing. 2019 56th ACM/IEEE design automation conference (DAC) (2019), pp. 1–6
54. B. Kahng, K. Samadi, CMP fill synthesis: A survey of recent studies. IEEE TCAD **27**(1), 3–19 (2018)
55. P. Liu, P. Tu, H. Wu, Y. Tang et al., An effective chemical mechanical polishing filling approach. Proceedings of ISVLSI (2015), pp. 44–49

56. Y. Bo, S. Sriraaman, ICCAD-2018 CAD contest in timing-aware fill insertion. Proceedings of ICCAD (2018), pp. 1–4
57. C. Feng, H. Zhou, C. Yan, J. Tao, et al., Efficient approximation algorithms for chemical mechanical polishing dummy fill. IEEE TCAD **30**(3), 402–415 (2011)
58. P. Gupta, A.B. Kahng, O.S. Nakagawa, K. Samadi, Closing the loop in interconnect analyses and optimization: CMP fill, lithography and timing. Proceedings of VMIC (2005), pp. 352–363
59. B. Kahng, K. Samadi, P. Sharma, Study of floating fill impact on interconnect capacitance. Proceedings of ISQED (2006), pp. 1–6
60. Dummy Filling Methods for Reducing Interconnect Capacitance and Number of Fills / T. Kurokawa, T. Kanamoto, A. Ibe, C.W. Kasebe, et al // Proc. ISQED (2005), pp. 586–591.
61. IBM Inc, CPLEX: High-performance mathematical programming solver for linear programming, mixed integer programming, and quadratic programming, Version 12.70. https://www.ibm.com/analytics/cplex-optimizer
62. R.O. Topaloglu, ICCAD-2014 CAD contest in design for manufacturability flow for advanced semiconductor nodes and benchmark suite. Proceedings of ICCAD (2014), pp. 367–368
63. S.S. Abazyan, V.A. Janpoladov, N.E. Mamikonyan, Standard cell pin access checking integration into test design verification. Proc. RA NAS NPUA. Ser. Tech. Sci. **73**(1), 74–81 (2020)
64. S.A.A.B. Olivier Aupoix, Optimizing standard cell pin accessibility in 14nm FDSOI with synopsys pin access checker. Synopsys Users Group (SNUG) (2014), pp. 1–6
65. I. Ricci, de Munari, P. Ciampolini, An evolutionary approach for standard-cell library reduction. Proceedings of ACM great lakes symposium on VLSI (GLSVLSI) (2007), pp. 305–310
66. W. Agatstein, K. McFaul, P. Themins, Validating an ASIC standard cell library. Proceedings of IEEE ASIC seminar and exhibit (1990), pp. 12/(6.1–6.5)
67. V. Melikyan, M. Martirosyan, A. Melikyan, G. Piliposyan, 14nm educational design kit: Capabilities deployment and future. Small Syst. Simul. Symp. (9), 1–5 (2018)
68. V.A. Janpoladov, A.A. Petrosyan, S.S. Abazyan, H.V. Margaryan, Random faults injection and simulation in auto-correction circuits. Proc. RA NAS NPUA Ser. Tech. Sci. **73**(2), 171–180 (2020)
69. S. Abazyan, V. Melikyan, Enhanced pin-access prediction and design optimization with machine learning integration. Microelectr. J. **116**, 1–5, 105198 (2021). https://doi.org/10.1016/j.mejo.2021.105198
70. S.-O. Shim, Multi-class classification based on relative distribution of class. 2020 2nd international conference on computer and information sciences (ICCIS) (2020), pp. 1–4. https://doi.org/10.1109/ICCIS49240.2020.9257679
71. P. Del Moral, S. Nowaczyk, S. Pashami, Hierarchical multi-class classification for fault diagnosis. Proceedings of the 31st European safety and reliability conference (2021), pp. 2457–2464. https://doi.org/10.3850/978-981-18-2016-8_524-cd.
72. S. Abazyan, Standard cell library enhancement for mixed multi-height cell design implementation. 2021 IEEE East-West design & test symposium (EWDTS) (2021), pp. 86–89. https://doi.org/10.1109/EWDTS52692.2021.9581045
73. C. Han, A. Kahng, L. Wang, B. Xu, Enhanced optimal multi-row detailed placement for neighbor diffusion effect mitigation in sub-10nm VLSI. IEEE Trans. Comput.-Aided Des. Integr. Circuits Syst. **PP**, 1–2 (2018). https://doi.org/10.1109/TCAD.2018.2859266
74. T. Lin, C. Chu, J. Shinnerl, Bustany, et al., POLAR: A high performance mixed-size wirelengh-driven placer with density constraints. IEEE Trans. Comput.-Aided Des. Integr. Circuits Syst. (34), 447–459 (2015). https://doi.org/10.1109/TCAD.2015.2394383
75. U. Brenner, BONNPLACE legalization: Minimizing movement by iterative augmentation. IEEE Trans. Comput.-Aided Des. Integr. Circuits Syst. **32**, 1215–1227 (2013). https://doi.org/10.1109/TCAD.2013.2253834
76. S.S. Abazyan, Method of designing power supply and grounding network of integrated circuits. Manual Natl. Acad. Sci. Repub. Armenia Natl. Polytech. Univ. Armenia. Tech. Sci. Ser. **74**(2), 197–203 (2021) (in Armenian)

77. M. Danigno, P. Butzen, J. Ferreira, A. Oliveira et al., Proposal and evaluation of pin access algorithms for detailed routing. 2019 26th IEEE international conference on electronics, circuits and systems (ICECS) (2019), pp. 602–605
78. K. Khalil, O. Eldash, A. Kumar, M. Bayoumi, Economic LSTM approach for recurrent neural networks. IEEE Trans. Circuits Syst. II Express Briefs **66**(11), 1885–1889 (2019). https://doi.org/10.1109/TCSII.2019.2924663
79. J. Verbraeken, M. Wolting, J. Katzy, J. Kloppenburg, et al., A survey on distributed machine learning. ACM Comput. Surv. **53**(2), 1–33 (2020). https://doi.org/10.1145/3377454
80. K. Khalil, O. Eldash, A. Kumar, M. Bayoumi, Machine learning-based approach for hardware faults prediction. IEEE Trans. Circuits Syst. I Regular Papers **67**(11), 3880–3892 (2020). https://doi.org/10.1109/TCSI.2020.3010743
81. Y. Du, Q. Ma, H. Song et al., Spacer-is-sielectric-compliant detailed routing for self-aligned double patterning lithography. Design automation conference (DAC), 2013 50th ACM/EDAC/IEEE (2013), pp. 1–6. https://doi.org/10.1145/2463209.2488848
82. C.-K. Cheng, D. Lee, D. Park, Standard-cell scaling framework with guaranteed pin-accessibility. 2020 IEEE international symposium on circuits and systems (ISCAS) (2020), pp. 1–5. https://doi.org/10.1109/ISCAS45731.2020.9180592
83. W.-T. J. Chan, Y. Du, A.B. Kahng et al., Beol stack-aware routability prediction from placement using data mining techniques. 2016 IEEE 34th international conference on computer design (ICCD) (2016), pp. 41–48. https://doi.org/10.1109/ICCD.2016.7753259
84. J.-R. Yu, D.D. Gao, et al., Accurate lithography hotspot detection based on principal component analysis-support vector machine classifier with hierarchical data clustering. J. Micro/Nanolithogr. MEMS MOEMS **14**(1), 1–12 (2014). https://doi.org/10.1117/1.JMM.14.1.011003
85. W.-T.J. Chan, P.-H. Ho et al., Routability optimization for industrial designs at sub-14nm process nodes using machine learning. Proceedings of the 2017 ACM on international symposium on physical design, ISPD '17 (Association for Computing Machinery, New York, 2017), pp. 15–21. https://doi.org/10.1145/3036669.3036681
86. R.-B. Lin, Y.-X. Chiang, Impact of double-row height standard cells on placement and routing (2019), pp. 317–322. https://doi.org/10.1109/ISQED.2019.8697712
87. J. Jooyeon, K.Taewhan, Utilizing middle-of-line resource in filler cells for fixing routing failures. 2021 IEEE international midwest symposium on circuits and systems (MWSCAS) (2021), pp. 1–5
88. S. Abazyan, Sh. Melikyan, D. Musayelyan, Standard cell library enhancement using neural network based sleep mode control integration for low leakage designs. 2021 IEEE East-West design & test symposium (EWDTS) (2021), pp. 105–108
89. S.A. Vitale, P.W. Wyatt, N. Checka, et al., FDSOI process technology for subthreshold-operation ultralow-power electronics. Proc. IEEE **98**(2), 333–342 (2010). https://doi.org/10.1109/JPROC.2009.2034476
90. M.-C. Kim, N. Viswanathan, Z. Li, C.J. Alpert, ICCAD-2013 CAD contest in placement finishing and benchmark suite. IEEE/ACM international conference on computer-aided design, digest of technical papers (2013), pp. 268–270. https://doi.org/10.1109/ICCAD.2013.6691130
91. S. Sreevidya, R. Holla, R. Raghu, Low power physical design and verification in 16nm FinFET technology. 2019 3th international conference on electronics, communication and aerospace technology (ICECA) (2019), pp. 936–940
92. A. Okazaki, VLSI researches for machine learning and neuromorphic computing. 2019 international symposium on VLSI technology, systems and application (VLSI-TSA) (2019), p. 1. https://doi.org/10.1109/VLSI-TSA.2019.8804628
93. M. Bartík, External power gating technique – An inappropriate solution for low power devices. 2020 11th IEEE annual information technology, electronics and mobile communication conference (IEMCON) (2020), pp. 0241–0245

94. A. Suren, M. Shavarsh, Educational open SPICE models neural network-based generation method. Proceedings of the 9th small systems simulation symposium (2022), pp. 50–54
95. Z. Xie, X. Xu, J. Hu, Y. Chen, Fast IR drop estimation with machine learning. Proceedings of the 39th international conference on computer-aided design (2020), pp. 1–8
96. Y. Chen, A.B. Kahng, G. Robins, A. Zelikovsky, Closing the smoothness and uniformity gap in area fill synthesis. Proc. ISPD, 137–142 (2002)
97. T. Ruiqi, D.F. Wong, R. Boone, Model-based dummy feature placement for oxide chemical-mechanical polishing manufacturability. IEEE Trans. Comput.-Aided Des. Integr. Circuits Syst. **20**(7), 902–910 (2001)
98. N. Mamikonyan, DRAM structure with prioritized memory bank using multi-VT bit cells architecture. 2020 IEEE East-West design & test symposium (EWDTS) (Varna, 2020), pp. 1–4. https://doi.org/10.1109/EWDTS50664.2020.9224821
99. J. Sercu, H. Barnes, Thermal aware IR drop using mesh conforming electro-thermal co-analysis. 2017 IEEE 21st workshop on signal and power integrity (SPI) (Baveno, 2017), pp. 1–4. https://doi.org/10.1109/SaPIW.2017.7944013
100. T. Lan et al., Timing-aware fill insertions with design-rule and density constraints. 2019 IEEE/ACM international conference on computer-aided design (ICCAD) (Westminster, 2019), pp. 1–8. https://doi.org/10.1109/ICCAD45719.2019.8942079

Index

A
Aging, 60–63, 76, 81–85, 87, 88, 101, 102
Analog-to-digital (ADC), 29, 130, 165
Asymmetries of rise/fall times, 144, 148

C
Calibration systems, 113, 121–123, 127–130, 132, 133, 135–137, 144–146, 148–150, 157
Characteristics, 3, 5, 8–12, 16, 17, 22, 27–34, 60, 68, 69, 73, 79, 99, 128, 159, 166–168, 171, 177, 179–183, 205, 230, 279, 281, 282, 285, 293, 296, 298, 299, 304, 306–329
Clock domains, 229, 231, 235, 240–242, 247, 248, 254, 256, 259, 268, 274
Comparators, 11, 20, 22, 24–26, 40–44, 46, 49, 52, 61–66, 76, 77, 81–82, 84–88, 102, 117, 145, 147, 148, 170–172, 177, 183, 185, 187, 191, 194–198, 204, 256, 257, 259, 263
Continuous time linear equalizer (CTLE), 12–18, 21, 22, 24–26, 28–39, 41–43, 46, 51, 52
Current sources, 15, 28, 38, 64, 70, 79, 84, 88, 90, 173–176, 183–187, 194, 196, 199, 205–207

D
Data transfer, 1, 2, 25, 27, 28, 49, 50, 52, 59, 67, 109, 111–113, 119, 121, 123, 126–128, 131, 137, 138, 145, 154, 223, 229–232, 235, 238, 240–243, 247, 250, 274
Decision Feedback Equalizer (DFE), 17–22, 26, 33, 40–45, 52
Deep learning (DL), 113, 115, 282, 284, 290, 291, 322–324
Design methods, 82, 83, 316
Digital cells, 281, 282, 296, 304, 306, 307
Digital delay lines (DDL), 63, 73–76, 80–82, 93, 97, 98, 100–102
Digital-to-analog converter (DAC), 21, 70, 117, 165
Direct memory access, 264
Dummy fill, 296
Duty cycle deviations, 119, 129, 131, 133, 138, 144

E
Equalizer, 8, 68

F
Flash ADCs, 170–172, 176, 177, 183–185, 194–196, 199, 201, 202, 204
Fluctuations, 73, 87, 93, 99, 102, 117, 144

H
Hidden layers, 291, 323
High frequency signal, 145, 159
High-performance heterogenous integrated circuits, 221–274
High-speed circuits, 129–137

I

IC design flow, 329
IC reliability, 59, 83, 124
Instruction sets, 223, 228, 243
Integrated circuits (ICs), 1–52, 59–102, 109–159, 165, 221–274, 279–329
I/O cell, 1–5, 9, 10, 20, 109–115, 117, 119, 121–123, 126–130, 137, 144, 145, 150, 279–286, 308–313, 315, 316, 323, 329

M

Machine learning (ML), 228, 282, 283, 285, 292, 295, 296, 304, 305, 308, 312–314, 316, 318, 319, 329
Metal fillers, 303, 306, 327, 329

N

Negative feedback, 46, 67, 82–83, 93, 100, 102, 189, 190, 199
Neural networks, 284, 291, 293, 296, 322, 323, 329
Non-standard operating conditions, 59, 76, 77, 81–83, 101

O

Offset, 22, 26, 41, 42, 46, 60, 125, 171
Offset errors, 172, 177, 178, 183, 185, 191, 194, 195, 197, 204, 210, 211, 214

P

Pipeline ADCs, 176–179, 189–192, 194, 210, 211, 214
Pulse Amplitude Modulation (PAM4), 8, 22, 25, 27, 46–52

S

Self-calibration, 159, 169, 170, 172, 173, 176, 183–188, 190–195, 198, 202, 207, 211, 214, 243, 244, 246
Signal distortions, 5, 9, 12, 69, 112, 116, 122, 137, 151, 159
Signal linearity, 165–194
Signal transmission, 2, 7, 8, 12, 22, 46, 113, 121, 122, 127–129, 148, 150, 155, 159
Speed of receiving sequential information, 25, 27
Stability, 33–35, 63, 76, 77, 80, 81, 83, 84, 201
Standard cells, 279, 281, 289, 296, 329

T

Transfer of information, 1–52

Printed in the United States
by Baker & Taylor Publisher Services